Synthesis Lectures on Mechanical Engineering

This series publishes short books in mechanical engineering (ME), the engineering branch that combines engineering, physics and mathematics principles with materials science to design, analyze, manufacture, and maintain mechanical systems. It involves the production and usage of heat and mechanical power for the design, production and operation of machines and tools. This series publishes within all areas of ME and follows the ASME technical division categories.

Tetsuya Ohashi

Mathematical Modeling of Dislocation Behavior and Its Application to Crystal Plasticity Analysis

Tetsuya Ohashi
Kitami Institute of Technology
Kitami, Japan

ISSN 2573-3168 ISSN 2573-3176 (electronic)
Synthesis Lectures on Mechanical Engineering
ISBN 978-3-031-37895-9 ISBN 978-3-031-37893-5 (eBook)
https://doi.org/10.1007/978-3-031-37893-5

© The Editor(s) (if applicable) and The Author(s), under exclusive license to Springer Nature Switzerland AG 2024

This work is subject to copyright. All rights are solely and exclusively licensed by the Publisher, whether the whole or part of the material is concerned, specifically the rights of translation, reprinting, reuse of illustrations, recitation, broadcasting, reproduction on microfilms or in any other physical way, and transmission or information storage and retrieval, electronic adaptation, computer software, or by similar or dissimilar methodology now known or hereafter developed.
The use of general descriptive names, registered names, trademarks, service marks, etc. in this publication does not imply, even in the absence of a specific statement, that such names are exempt from the relevant protective laws and regulations and therefore free for general use.
The publisher, the authors, and the editors are safe to assume that the advice and information in this book are believed to be true and accurate at the date of publication. Neither the publisher nor the authors or the editors give a warranty, expressed or implied, with respect to the material contained herein or for any errors or omissions that may have been made. The publisher remains neutral with regard to jurisdictional claims in published maps and institutional affiliations.

This Springer imprint is published by the registered company Springer Nature Switzerland AG
The registered company address is: Gewerbestrasse 11, 6330 Cham, Switzerland

Preface

Plastic deformation of metallic materials is a matter of interest in diverse areas of science, technology, everyday life, and human history. This book is dedicated to describe physical background and its mathematical modeling of plastic slip behavior observed in the microstructure of metal crystals. Lattice defects play major roles during the plastic slip deformation, and among those, behavior of dislocations is significant. How do dislocations move under load in the environment with internal boundaries or different phases? How do macroscopic properties such as yield and strain hardening occur and why do they depend on microscopic length scales, such as grain diameter or precipitate spacing? How can these phenomena be modeled by mathematical equations and used in numerical analysis techniques such as finite element method? This book is written for students, scientists, and engineers involved in materials science, mechanical engineering, and microelectronics.

Kitami, Japan   Tetsuya Ohashi

Acknowledgements

Dr. Yelm Okuyama of Kisarazu National College of Technology and Dr. Yohei Yasuda of Asahikawa National College of Technology cooperated in the revision of the software for crystal plasticity analysis described in this book. Dr. Tsuyoshi Mayama of Kumamoto University provided valuable information, especially on the deformation characteristics of HCP crystals. Dr. Masaki Tanaka of Kyushu University provided materials on the microstructural characteristics near the fracture surface of materials and theoretical considerations on the thermal activation process of deformation. Dr. Hussein Zbib of Washington State University provided dislocation dynamics simulation software code and we discussed various research topics. I am especially grateful to Dr. Kazuyuki Shizawa of Keio University, Dr. Kenji Higashida of Kyushu University, Dr. Kaneaki Tsuzaki of National Institute for Materials Science, Dr. Tomotsugu Shimokawa of Kanazawa University, and many colleagues and friends for giving me various opportunities to discuss and think. This research was partly supported by the following research funds: MEXT Japan under Grant No. 18062001 in Priority Areas "Giant Straining Process"; The Japan Science and Technology Agency (JST) in Collaborative Research Based on Industrial Demand "Heterogeneous Structure Control, Towards Innovative Development of Metallic Structural Materials"; MEXT Japan, Grant No. 23109001 and 23109008 in Grant-in-aid for Scientific Research on Innovative Areas, "Materials Science on Synchronized LPSO Structure"; the Council for Science, Technology and Innovation (CSTI), Cross-ministerial Strategic Innovation Promotion Program (SIP), "Structural Materials for Innovation", (Funding Agency, Japan Science and Technology Agency). I would like to express my gratitude to all these people and agencies.

Contents

1. **Introduction** 1
   References 2
2. **Plastic Shear Strain and Dislocation Movement** 3
3. **Dislocation Accumulation Due to Plastic Slip** 7
   3.1 Statistically Stored Dislocations 7
     3.1.1 Plastic Shear Strain and Accumulation of Dislocations Due to Emission of Dislocation Loops 7
     3.1.2 Strain Hardening Associated with the Evolution of Dislocation Density 10
   3.2 Geometrically Necessary Dislocations 18
   References 23
4. **Models for the Critical Resolved Shear Stress** 25
   4.1 Dislocation Accumulation and Strain Hardening 26
     4.1.1 One-Dimensional Approach 26
     4.1.2 Extension to Three-Dimensional Space and Inclusion of the Effect of Twin Deformation 28
     4.1.3 Deformation Modes in FCC, BCC and HCP Crystals and Their Interaction Matrix 29
     4.1.4 An Extended Model for the Dislocation Mean Free Path 39
   4.2 Effect of Length Scale of Metal Microstructure on Strength and Strain Hardening 40
     4.2.1 Geometrically Necessary Dislocations and Scale Dependent Strain Hardening Characteristics 40
     4.2.2 Scale Dependent Modeling of the Yield Stress of Polycrystals 41
     4.2.3 Deformation Twin and Critical Resolved Shear Stress 49
     4.2.4 Introduction of Microstructural Length Scale into the Model of the Dislocation Mean Free Path 51
   References 55

5 Generation of Atomic Vacancies by Dislocation Pair Annihilation 57
  References 66

6 Mathematical Framework of Crystal Plasticity Analysis Constructed
  by Hill and Implementation of Dislocation Models 67
  Reference 68

7 An Example of Analysis: Tension–Compression Straining of a FCC
  Single Crystal Plate and Bauschinger Effect 69
  References 76

**Appendix A: Coordinate Transformation by Euler Angles and Graphical
Presentation by Pole Figures** 77

Appendix B: Work Done by Slip Deformation and Equivalent Plastic Strain ... 81

Index 83

Introduction 1

Abstract

The study of plastic deformation of metals requires both the science of solid mechanics and the materials science of lattice defects in crystals, and mathematical modelling of the physical elementary processes of plastic deformation serves as a link between the two. In this chapter, we briefly review the history of this research until now and indicate the direction of this book.

Keywords

Yielding and strain hardening • Characteristic length scale • Microstructure

The study of plastic deformation of metallic materials has two distinct aspects: the deformation mechanics of crystalline solids and the physical properties of the material that cause the phenomenon. In 1982, Peirce et al. [1] reported the results of an crystal plasticity analysis showing that when tensile load is applied to a two-dimensional single-crystal plate, shear bands suddenly forms in the specimen and the slope of the macroscopic load-elongation curve becomes negative. The paper incorporates features such as the representation of the large deformation with anisotropy of crystal slip deformation, the evaluation of the crystal rotation during deformation, and the representation of the transition from the elastic to elastoplastic state as a continuous mathematical function, which are integrated into the framework of the finite element method. The main focus of this study was on the solid-mechanics aspects of plastic deformation of crystalline materials.

On the other hand, there is a long history of research from a materials science perspective [2, 3], and the important keywords that link the solid mechanics and the materials science are material yielding and strain hardening. Yielding, the onset of plastic deformation, is caused by the motion of dislocations in the material. The strain hardening associated with plastic deformation is also described by dislocations. The reason why macroscopic yield stress and strain hardening are linked to the characteristic length scale of the material's microstructure (e.g., average grain size) are also understood by the behavior of dislocations. However, the link between the mechanics of deformation of solids and materials science is not yet well established.

Deformation in the microstructure of metal crystals is a physical phenomenon that spans multiple scales. Focusing on deformation phenomena at the grain level, there is a strong anisotropy in deformation properties of crystals and phases with different mechanical properties are present in the microstructure. Therefore, macroscopically uniform stress field is disturbed by the interfaces in the microstructure, resulting in the non-uniform slip deformation within crystal grains and multiplication of slip on different slip systems. Under such an environment, dislocations interact with solute atoms, atomic vacancies, accumulated dislocations, precipitates, grain- or phase boundaries and so on. The driving force for the movement of dislocations is the non-uniformly distributed stress field and numerical analysis approaches such as finite element method is essential as a means of evaluating the mechanical field.

A number of mathematical models have been established to describe the collective behavior of dislocations. However, many of them have not been developed to a form suitable for three-dimensional numerical analysis. In this book, we briefly review the background of these models, quantify and expand them to three-dimensional space. We also introduce some examples of application and research subjects to be considered in the future.

References

1. Peirce D, Asaro R, Needleman A (1982) Acta Metall 30:1087
2. Cottrell AH (1952) Dislocations and plastic flow in crystals, 1st editio. Oxford University Press, Oxford
3. Kocks U, Argon A, Ashby M (1975) Prog Mater Sci 19:1

Plastic Shear Strain and Dislocation Movement

Abstract

Some basic concepts in plastic slip deformation of crystalline materials, movement of dislocations, and plastic shear strain realized by the dislocation movement are briefly given.

Keywords

Plastic slip deformation · Crystalline materials · Movement of dislocations · Plastic shear strain

Let us summarize the plastic deformation due to the movement of straight or looped dislocations in an infinite or not-constrained medium. First, assume that a rectangular specimen of size $w \times h \times d$ in x, y and z direction with a straight dislocation of edge type lying in z direction with Burgers vector parallel to x axis. As shown in Fig. 2.1a, when the dislocation path through the specimen cross section under an application of shear stress, a plastic displacement equal to the magnitude b of the Burgers vector **b** occurs. If the distance of movement of the dislocation is smaller than w, plastic displacement that occurs in the specimen can be approximated by $b \times (x/w)$. If n dislocations move as shown in Fig. 2.1b and $\bar{x}$ is the average of their travel distance, the plastic displacement is approximated by $nb\bar{x}/w$ and the plastic shear strain of the specimen is given by,

$$\gamma = nb\bar{x}/wh. \qquad (2.1)$$

Density of dislocations is given by the total dislocation length divided by the volume and therefore, $n/wh = nd/whd$ is equal to the density. Denoting the density by ρ, we have the relation,

$$\gamma = \rho b \bar{x}. \qquad (2.2)$$

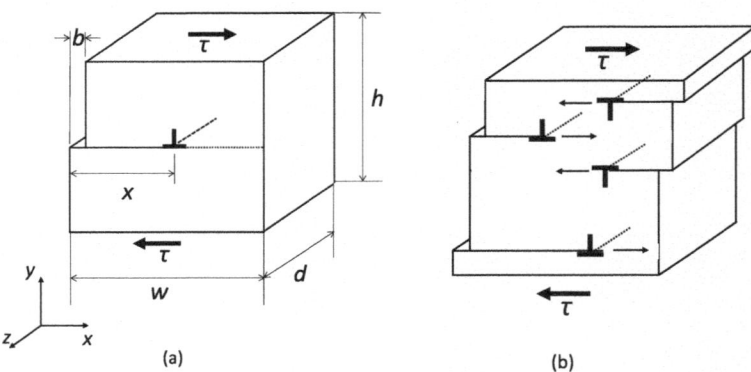

Fig. 2.1 **a** Movement of a dislocation line results in plastic displacement of magnitude b, **b** when multiple dislocation line move, each contribute to the plastic displacement in a linear sum

Differentiating the shear strain with time yields,

$$\dot{\gamma} = \rho b \bar{v}, \tag{2.3}$$

where, $\bar{v}$ is the average velocity of moving dislocations. Experimental observations have shown that,

$$v = A\tau^m, m \approx 1, \tag{2.4}$$

where τ is the applied stress. Equation (2.4) is a base for dislocation dynamics simulations.

Dislocations have an apparent line tension because they involve elastic strain energy. When a shear stress is applied to a dislocation segment pinned at both ends, the dislocation moves on the slip plane with a curved shape as shown in Fig. 2.2a. Figure 2.2b–f show results of dislocation dynamics simulation of this process. At the stage of (e), parts m and n of the dislocation line meet. Because they have the same Burgers vector and opposite line direction, they annihilate each other and a dislocation segment connecting the two pinning points and a closed loop are left. This process is called the Frank-Read type dislocation loop emission. If this process takes place in an infinite medium and the shear stress continues to be applied, emitted dislocation loop expand to distant locations and the dislocation segment continues to emit dislocation loops repeatedly as similar as those in Fig. 2.2c–f. The pinned dislocation segment is called the Frank-Read type dislocation source. Similar process takes place even when the dimension of the specimen is finite if the emitted dislocation lines escape from the specimen surfaces.

The shear stress needed to emit dislocation loops is equal to the stress needed to overcome the stage when the curvature of the dislocation line is the maximum shown in (c) and is called the Orowan stress,

Fig. 2.2 **a** A dislocation segment AB is assumed to lie on a slip plane and points A and B are pinned. When shear stress is applied, the segment bows out on the slip plane. **b** The initial state, **c–f** dislocation dynamics simulation results for the process of the dislocation movement. When the parts m and n in **e** of the dislocation line meet, they annihilate and a closed loop is formed, leaving the segment between A and B. The minimum stress needed to accomplish this process is approximately given by μb divided by the distance λ between A and B and the segment AB is called the Frank-Read source

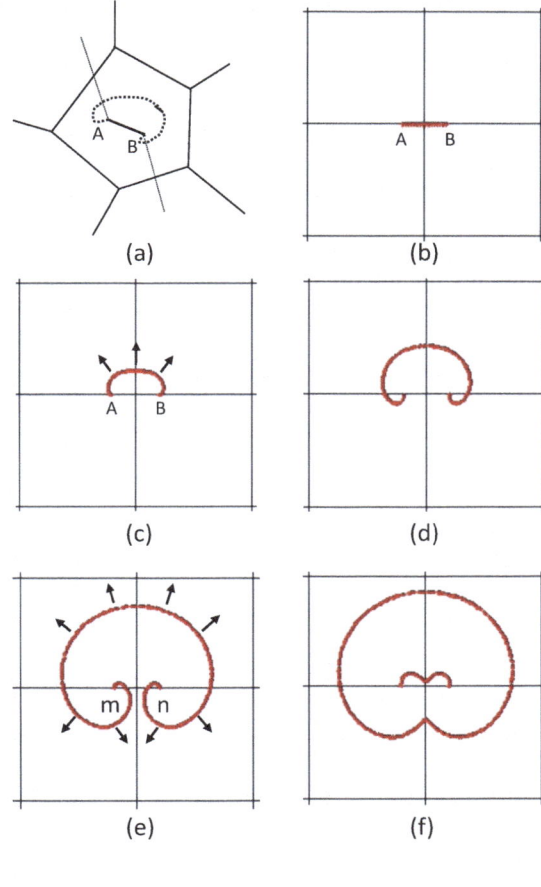

$$\tau_\infty \approx \frac{\mu b}{\lambda}. \tag{2.5}$$

where, μ and λ denote the elastic shear modulus and the distance between points A and B, respectively.

If another obstacle other than the two pins to its movement is encountered during the expansion of the dislocations arc, this makes another cause of increase in the curvature and the shear stress needs to be increased. This happens when dislocation loops expand in grains with dispersed hard particles or inside grains of small dimensions. Details for these processes will be discussed in Chap. 4.

Dislocation Accumulation Due to Plastic Slip 3

Abstract

First, the accumulation and annihilation of statistically stored dislocations due to emission of dislocation loops is described and formulated. The important concepts of dislocation mean free path and annihilation distance are introduced and strain hardening associated with slip deformation are shown for some cases with simplified parameter sets. Next, the spatial gradient of plastic shear strain and geometrically necessary dislocations are explained in detail and extended to three-dimensional space. The direction vectors and characteristic angles of the geometrically necessary dislocations in 3-D space are derived and used to visualize the emission of prismatic dislocation loops from a micrometer-sized void.

Keywords

Slip deformation · Movement of dislocations · Plastic shear strain · Dislocation mean free path · Dislocation annihilation · Strain hardening · Critical resolved shear stress · Taylor model · Geometrically necessary dislocations · Gradient of plastic shear strain · Prismatic dislocation loops

3.1 Statistically Stored Dislocations

3.1.1 Plastic Shear Strain and Accumulation of Dislocations Due to Emission of Dislocation Loops

Let us consider the increase of dislocation density and plastic strain due to emission of dislocation loops. When a dislocation loop completely sweeps the cross section of the specimen, the resultant plastic shear displacement is equal to b and this is similar to the case when straight dislocation line cut the cross section. If n dislocation loops sweep

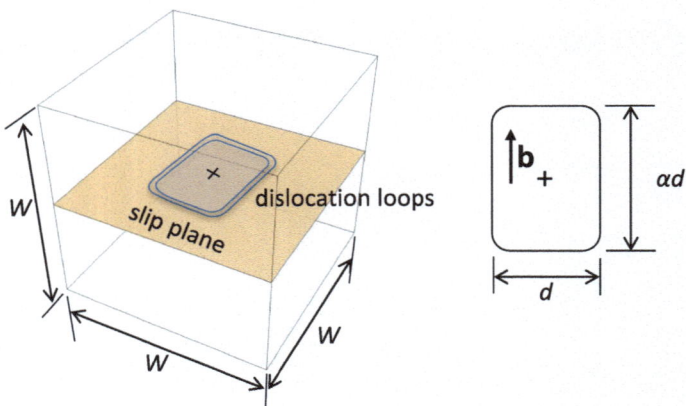

Fig. 3.1 Assuming that rectangular dislocation loops of size $d \times \alpha d$ are generated and stopped, the increments of plastic shear strain and dislocation density are evaluated

the specimen cross section, the resultant plastic displacement is nb. If the movement of dislocation loops is stopped before the complete sweep, something similar to what we considered to derive Eq. (2.1) occurs. When the expansion of dislocation loops is stopped at a rectangle of $d \times \alpha d$, as shown in Fig. 3.1, the total area swept by n dislocation loops is,

$$\Delta A = n \times \alpha d^2, \qquad (3.1)$$

and the plastic shear strain produced by this expansion is proportional to ΔA divided by the cross-sectional area W^2,

$$\Delta \gamma = \frac{b}{W} \cdot \frac{\Delta A}{W^2}. \qquad (3.2)$$

Since the loop has side lengths d and αd, travel distances of dislocation segments are $d/2$ and $\alpha d/2$, and the average distance moved is,

$$L = \left(\frac{d}{2} + \frac{\alpha d}{2} \right) \bigg/ 2 = \frac{d(1+\alpha)}{4}. \qquad (3.3)$$

L is called the dislocation mean free path. The total length of accumulated dislocations is $n(2\alpha d + 2d)$ and the increment of dislocation density $\Delta \rho$ is,

$$\Delta \rho = n(2\alpha d + 2d)/W^3. \qquad (3.4)$$

Using Eqs. (3.1)–(3.4), we obtain

3.1 Statistically Stored Dislocations

$$\Delta\rho = \frac{(1+\alpha)^2}{2\alpha} \cdot \frac{1}{bL} \cdot \Delta\gamma, \quad (3.5)$$

or in differential form,

$$d\rho_S^+ = \frac{(1+\alpha)^2}{2\alpha} \cdot \frac{1}{bL} \cdot d\gamma. \quad (3.6)$$

The dislocations described above exists in pairs with positive and negative signs of the Burgers vector. Dislocations where the increment of dislocation density is proportional to the increment of plastic shear strain and the sum of Burgers vectors of the dislocations accumulated in a small region is zero is called statistically stored dislocations, or SS dislocations.

Since there are equal numbers of dislocations with positive and negative signs in the Burgers vector of SS dislocations, if dislocations with different signs are located in close proximity, they will be annihilated by the thermal activation process as schematically shown in Fig. 3.2. Assuming that the amount of annihilation is proportional to the dislocation density and plastic shear strain, the amount of reduction can be written as follows,

$$d\rho_S^- = -D \cdot \rho_S \cdot \frac{d\gamma}{b}. \quad (3.7)$$

The amount of dislocation annihilation given by Eq. (3.7) is linked to the amount of increase in plastic shear strain, which is called dynamic recovery. The coefficient D is called the annihilation distance and is assumed to be a function of strain rate, temperature, microstructural length scale, stress state, and so on. That is,

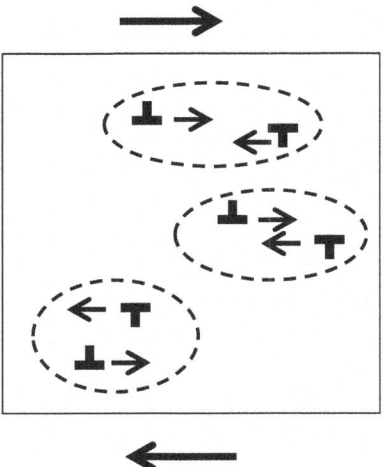

Fig. 3.2 As plastic deformation proceeds with the accumulation of positive and negative dislocations, the annihilation of dislocations occurs simultaneously. This process is called dynamic recovery

$$D = D(\dot{\gamma}, T, d, \sigma, \ldots). \tag{3.8}$$

The accumulation given by Eq. (3.6) and the annihilation given by Eq. (3.7) both contribute to the density evolution of SS dislocations and we obtain an important relation [1],

$$d\rho_S = \left(\frac{(1+\alpha)^2}{2\alpha} \cdot \frac{1}{bL} - \frac{D}{b}\rho_S \right) \cdot d\gamma. \tag{3.9}$$

In order to distinguish changes in the densities of SS dislocations in separate slip systems, Eq. (3.9) can be rewritten as,

$$d\rho_S^{(n)} = \frac{1}{b^{(n)}} \left(\frac{(1+\alpha)^2}{2\alpha} \cdot \frac{1}{L^{(n)}} - D^{(n)} \rho_S^{(n)} \right) \cdot d\gamma^{(n)}. \tag{3.10}$$

Here, superscript (n) indicates the slip system number.

3.1.2 Strain Hardening Associated with the Evolution of Dislocation Density

Scale dependent parameters of the dislocation mean free path L and the annihilation distance D take important roles in the accumulation of SS dislocations. Let us examine some simplified cases of dislocation accumulation in the light of strain hardening.

The shear stress that must be applied to produce plastic slip in a slip system is called critical resolved shear stress (CRSS). Let us denote the CRSS by θ. Generally, the CRSS is related to the density of accumulated dislocations by the Taylor relation,

$$\theta = a\mu b\sqrt{\rho_S}. \tag{3.11}$$

Details and its development of the CRSS will be discussed in the next section, however, we briefly examine the relationship between strain hardening and the two length scale-parameters L and D used in Eq. (3.9).

3.1.2.1 Stress–Strain Curves When the Dislocation Mean Free Path is a Constant

First, let us examine the case when the dislocation mean free path is a constant value of L_0. If $D = 0$, we have from Eq. (3.9),

$$\rho_S = \frac{c}{bL_0}\gamma + \rho_0, \tag{3.12}$$

where, ρ_0 denotes the initial dislocation density and,

3.1 Statistically Stored Dislocations

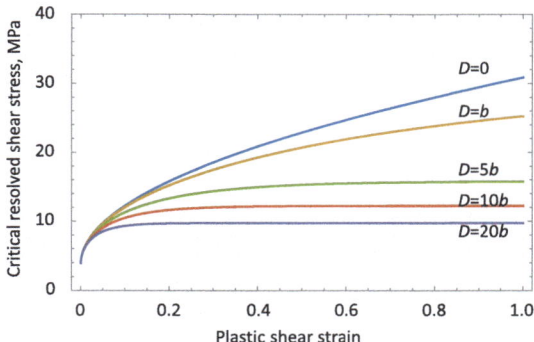

Fig. 3.3 Stress–strain curves when the dislocation mean free path is kept constant and the annihilation distance D is assumed to be 0 to $20b$. Dislocation density is evaluated by Eq. (3.9) and the plastic flow stress is given by the sum of Taylor term given in Eq. (3.11) and a friction stress of 3.7 MPa. Parameters used are, $a = 0.1$, $\mu = 3.85 \times 10^{10}$ Pa, $b = 2.5 \times 10^{-10}$ m, $c = 2$, $\theta_0 = 3.7$ MPa, $L_0 = 10$ μm, $\rho_0 = 1 \times 10^{11}$ m^{-2}

$$\frac{(1+\alpha)^2}{2\alpha} = c. \quad (3.13)$$

Applying Eq. (3.12) to the Taylor model of Eq. (3.11) yield a parabolic strain hardening property where the CRSS is proportional to the 1/2 power of plastic shear strain, which is usually called Stage I hardening.

Analytical integration of Eq. (3.9) is possible when $D > 0$, also, and the density of SS dislocations evolves with plastic shear strain as follows,

$$\rho_S = \frac{c}{DL_0} + \left(\rho_0 - \frac{c}{DL_0}\right)\exp\left(-\frac{D}{b}\gamma\right). \quad (3.14)$$

c/DL_0 is the asymptotic dislocation density when $\gamma \to \infty$. Applying Eq. (3.14) to the Taylor model of Eq. (3.11), we obtained strain hardening curves shown in Fig. 3.3. Here, $L_0 = 10$ μm, $\rho_0 = 10^{11}$ m^{-2} were used. It can be seen that the increase in CRSS becomes to be negligible when the plastic shear strain is about 0.2 for $D = 10b$ and 0.1 for $D = 20b$. Another set of parameters for the initial values of the dislocation mean free path and dislocation density ($L_0 = 1000$ μm, $\rho_0 = 10^9$ m^{-2}) was used to examine their effect, but the shear strain when the CRSS value saturate did not change largely.

3.1.2.2 Stress–Strain Curves When the Dislocation Mean Free Path Evolves with Deformation

As another simplified example, consider the case where the dislocation mean free path evolves with deformation under the condition $D = 0$. If we assume that the mean free path is inversely proportional to the strain,

$$L = \frac{\Lambda}{\gamma}, \tag{3.15}$$

then, the increase in dislocation density is,

$$d\rho_S = \frac{c}{bL} \cdot d\gamma = \frac{c\gamma}{b\Lambda} \cdot d\gamma. \tag{3.16}$$

Integrating Eq. (3.16), we obtain,

$$\rho_S = \frac{c}{b\Lambda} \cdot \frac{1}{2}\gamma^2 + \rho_0. \tag{3.17}$$

Applying Eq. (3.17) to the Taylor model of Eq. (3.11) yields the linear strain hardening curve typically observed in the deformation stage II of FCC single crystals.

Example 1 Model by Seeger et al.

When FCC crystals oriented for single slip are deformed by tensile load, mostly uniform activity of a slip system with the highest Schmid factor is observed at the initial period of deformation and the load-elongation curve is approximately parabolic and constitutes the stage I deformation curve. As deformation proceed, the load-elongation curve transforms to a linear one with a larger gradient and forms the stage II period. In the model of Seeger et al. [2], the dislocation mean free path is designed to reproduce this transition from stage I to II when the plastic shar strain reaches a predetermined value of $\gamma*$.

$$L = \begin{cases} L_0 \cdots \text{when } \gamma \leq \gamma* \\ \frac{\Lambda}{\gamma - \gamma*} \cdots \text{when } \gamma > \gamma* \end{cases}. \tag{3.18}$$

Because the mean free path is a constant L_0 for the period $\gamma \leq \gamma*$, dislocation density is proportional to the plastic shear strain and the strain hardening property becomes parabolic. When the deformation advances and $\gamma > \gamma*$, the dislocation density is proportional to the square of the plastic shear strain as shown in Eq. (3.17) and the strain hardening characteristics transforms to linear one. Equation (3.18) is a model that expresses transition from a parabolic strain hardening curve, or stage I deformation, to a linear strain hardening curve, or stage II deformation at a given strain $\gamma*$.

The model given by Eq. (3.18) has a minor flaw that the mean free path is discontinuous at $\gamma = \gamma*$. To avoid this, the model was modified as follows and is shown in Fig. 3.4,

$$L = \begin{cases} L_0 \cdots \text{when } \gamma \leq \gamma* \\ \frac{\Lambda}{\gamma - (\gamma* - \Lambda/L_0)} \cdots \text{when } \gamma > \gamma* \end{cases}. \tag{3.19}$$

3.1 Statistically Stored Dislocations

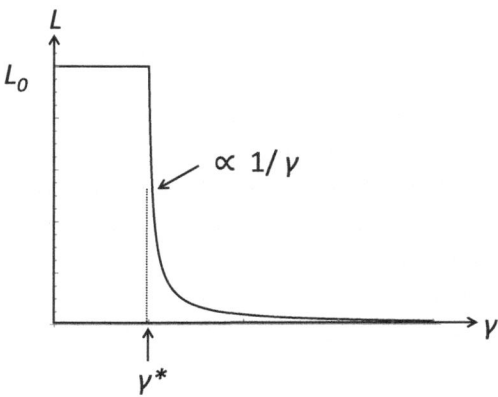

Fig. 3.4 Model of Seeger et al. [2] for the dislocation mean free path. The dislocation mean free path remains constant until the plastic shear strain reaches a given value γ^*, and after that, decreases in inversely proportional manner

Experimental data on dislocation mean free paths in the stage I deformation of pure copper single crystals are available [3–6] and found to be in the range of approximately 1000–5000 μm, depending on initial conditions and preparation. Figure 3.5 shows stress–strain relationship obtained by the model of Eq. (3.19) and the numerical results appear to agree with experimental results [3] well.

Example 2 Spontaneous transition from stage I to II in tensile deformation of FCC single crystals [7–10].

The model of Seeger et al. given by Eqs. (3.18) or (3.19) is designed so that the transition from stage I to II occurs when the plastic shear strain reaches a given value of γ^*. In reality however, the amount of deformation until the transition occurs varies with

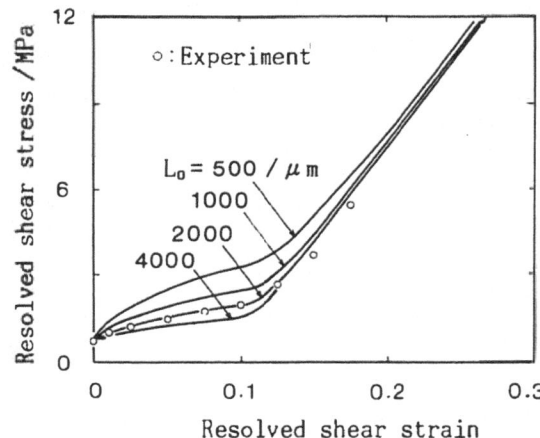

Fig. 3.5 Stress–strain curves for Cu single crystals obtained with the model for dislocation mean free path by Seeger et al. The initial dislocation density is $\rho_0 = 5 \times 10^8$ m^{-2} and $\gamma^* = 0.1$. Experimental data by Basinski and Basinski [3] are given by symbols ○ [7]

the initial crystal orientation, heat treatment, and other factors. Let us consider a model in which transition occurs spontaneously.

The transition from deformation stage I to II has been studied in detail by experimental observation. Higashida et al. [11] showed that fine activity of a slip system different from the primary system occurs sporadically in the specimen as a precursor of stage II, and at the same time, slip lines of the primary systems show kinks. Let us modify the dislocation mean free path model where two functions, the first is a constant and the second is inversely proportional to the shear strain, switch when a slip system other than the primary one activates and multiplied to the primary one,

$$L^{(n)} = \begin{cases} L_0^{(n)} \cdots single\ slip \\ \dfrac{\Lambda}{\sum_m \gamma^{(m)} - \left(\gamma^D - \dfrac{\Lambda}{L_0^{(n)}}\right)} \cdots multiple\ slip\ . \end{cases} \quad (3.20)$$

Here, n and m denote slip system numbers and if the crystal is FCC type and twelve slip systems of {111}<110> are considered, n, m = 1, 2, ..., 12. γ^D is the plastic shear strain on the primary system when the slip multiplication occur, as shown in Fig. 3.6. The amount of shear strain on the secondary system could be very small and at the order of 10^{-6}. No matter what sample preparation procedures are followed, there will be initial inhomogeneities in single crystal specimens [12], such as non-uniform distribution of grown-in dislocations, subtle spatial deviation of crystal orientation or fluctuations in the density distribution of impurity atoms. We assume such inhomogeneities and, during the analysis, we monitor the number of active slip systems at each point in the specimen. When multiple slip is detected, the function for the dislocation mean free path switches from the constant to inversely proportional one.

Single crystal specimens with spatially non-uniform initial dislocation density were prepared and their uniaxial tensile deformation was analyzed by a crystal plasticity finite element method [9, 10]. Rectangular shaped specimen was divided into $5 \times 5 \times 15$ finite elements and the initial dislocation densities in each element was given by a normal

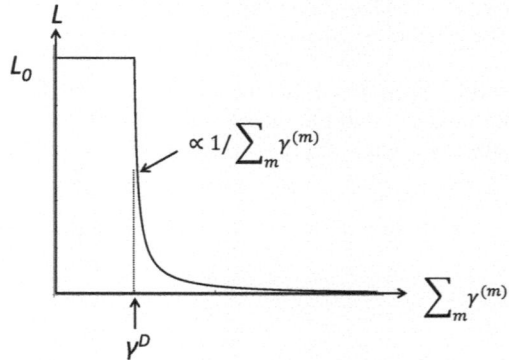

Fig. 3.6 Modified model of Seeger et al. for the dislocation mean free path. The dislocation mean free path is constant until the superposition of slip activity in the secondary slip system and decreases inversely after the superposition

3.1 Statistically Stored Dislocations

random numbers. The median and standard deviation of the initial density are 1×10^9 and 0.32×10^9 m^{-2}, respectively. The initial value of the dislocation mean free path, L_0 was 1000 μm and the coefficient $\Lambda = 4$ μm was used.

The slip deformation of the primary slip system was not uniform due to the initial inhomogeneity, and with this non-uniform deformation, the direction of principal stress gradually shifted away from the loading axis and tri-axial stress state was formed. This tri-axial stress state resulted in a superimposed secondary slip with shear strain of the order 10^{-5}. Load-elongation curves for specimens with different initial orientations are shown in Fig. 3.7. Specimens with its initial orientation well inside the stereo triangle showed longer duration of stage I deformation with parabolic strain hardening character, followed by the stage II with linear strain hardening. Duration of stage I varies with the initial crystal orientation; the closer to the double slip orientation, the shorter is the duration of stage I. Fine activation of the secondary slip system was observed during the transition from stage I to II. These results agree well with experimental ones [13].

Example 3 Model of the dislocation mean free path defined by state variables.

In the models examined in Examples 1 and 2, the dislocation mean free path was defined as a function of plastic shear strain. The plastic shear strain, however, cannot be a variable that objectively describes the internal state of the crystal, because the strain

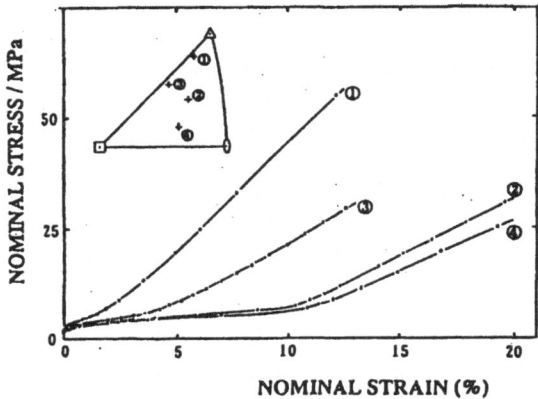

Fig. 3.7 Stress–strain curves for Cu single crystals obtained with the modified model of Seeger et al. for the dislocation mean free path. Rectangular specimen is divided into $5 \times 5 \times 15$ finite elements and the initial dislocation density in each elements is given by a normal random numbers. Median and standard deviation of the initial density are 1×10^9 and 0.32×10^9 m^{-2}, respectively. The initial value L_0 of the dislocation mean free path is 1000 μm and the coefficient $\Lambda = 4$ μm was used [10]

describes the deformation based on an artificially defined initial state. Density of dislocations, on the other hand, is an internal state variable with objectivity that evolves with deformation.

Let us assume that the dislocation mean free path is defined by the density of accumulated dislocations on slip systems and other length scale parameters related to the metal microstructure;

$$L^{(n)} = L^{(n)}\left(\rho_a^{(1)}, \rho_a^{(2)}, \cdots, d_1^{*(n)}, d_2^{*(n)}, \cdots\right). \tag{3.21}$$

$\rho_a^{(1)}, \rho_a^{(2)}, \cdots$ are the density of dislocations accumulated on slip systems 1, 2, ... and $d_1^{*(n)}, d_2^{*(n)}$. ... are various length scales defined in the microstructure. Possible modeling for the length scales will be discussed in Sects. 4.1.4 and 4.2.4, and we discuss here on the dislocation density and the mean free path.

Assuming that the average distance of accumulated dislocations defines the dislocation mean free path, it can be written as [14],

$$L^{(n)} = \frac{c^*}{\sqrt{\sum_m \rho_a^{(m)}}}. \tag{3.22}$$

Here, $1/\sqrt{\sum_m \rho_a^{(m)}}$ is the average spacing of accumulated dislocations and Eq. (3.22) models that the mean free path of dislocations moving on slip system n is given by c^* times the average distance of dislocations accumulated on slip systems. The value of c^* is related to the probability of trapping by interacting dislocations and is assumed [14] to be $c^* = 10$–100. An expansion of the Eq. (3.22) could be [9],

$$L^{(n)} = \frac{c^*}{\sqrt{\sum_m w^{(nm)} \rho_a^{(m)}}}, \tag{3.23}$$

where, a weight matrix $w^{(nm)}$ is introduced. The strength of the contribution of accumulated dislocations on slip system m to the mean free path of dislocations on slip system n is incorporated through the component of the weight matrix.

As will be discussed further in Sect. 4.1.3, there are various interactions between accumulated and moving dislocations, but the simplest classification is whether they cut each other or interact only elastically without mutual cutting. Dislocations belonging to the same slip system or sharing a slip plane (coplanar systems) interact only through elastic field. In this case, it may be possible to simplify that these dislocations do not contribute to each other's mean free path and this is expressed by placing zeros in the corresponding components of the weight matrix $w^{(nm)}$.

If we assume that the mean free path is determined by mutual cutting of dislocations, interaction will be a thermal process and possible form of c^* will be as follows [15],

3.1 Statistically Stored Dislocations

$$c^* = c_a^* + c_0^* exp\left(\frac{-G^*}{k_B T}\right), \quad (3.24)$$

where, k_B, T and G^* denote the Boltzmann constant, temperature and activation energy for the dislocation cutting, respectively, while, c_a^* and c_0^* are constants. The first and second terms on the right-hand side of Eq. (3.24) represent the athermal and thermal terms. The contribution of the thermal term largely depends on the constant c_0^* and the activation energy G^*. Kujirai and Shizawa [15] used $c_0^* = 1300$ and $G^* = 55$ zJ ($= 55 \times 10^{-21}$ J ≈ 33 kJ/mol) in the study of high temperature deformation of Ni polycrystals at $T = 1000$ K. Figure 3.8 shows c^* as a function of temperature and it can be seen that the contribution from the thermal activation is negligibly small near room temperature. Also shown is the change in c^* when $G^* = 20$ zJ (≈ 12 kJ/mol), where the contribution from the thermal activation is pronounced near room temperature.

Let us consider the case when the activation energy is large and c^* is approximated to be athermal and a constant. Assuming that components of the weight matrix for self-systems to be zero and otherwise chosen to satisfy quasi-isotropic condition, we analyzed tensile deformation of FCC single crystals with initial inhomogeneity in dislocation density [9, 10]. Results are shown in Fig. 3.9. In specimens oriented for single slip and shown by symbols □, the strain hardening show parabolic shape, while specimen oriented for double slip and shown by symbols ○ exhibit approximately linear hardening property and strain hardening ratios for both cases agree well with experimental results. On the other hand, spontaneous transition from stage I to II in specimens oriented for single slip was not clearly observed even when large elongational strain was applied. Similar to the

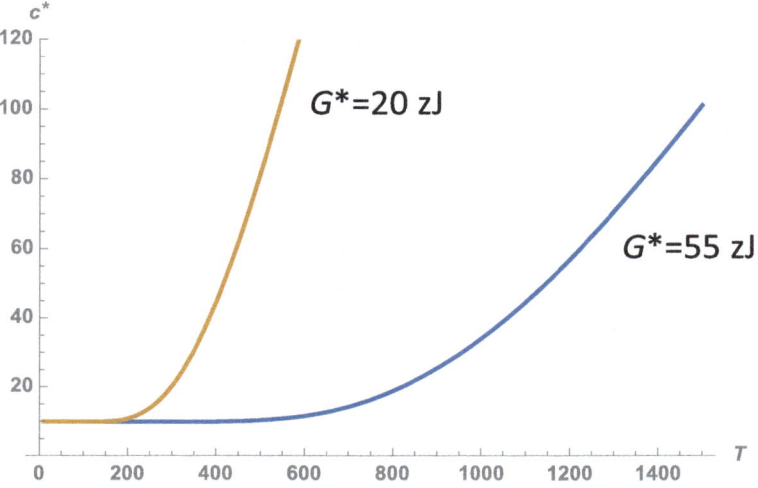

Fig. 3.8 A temperature dependent model [15] for the coefficient c^* in Eq. (3.23). c^* is given by athermal and thermal terms as shown in Eq. (3.24). $c_a = 10$, $c_0 = 1300$ are assumed

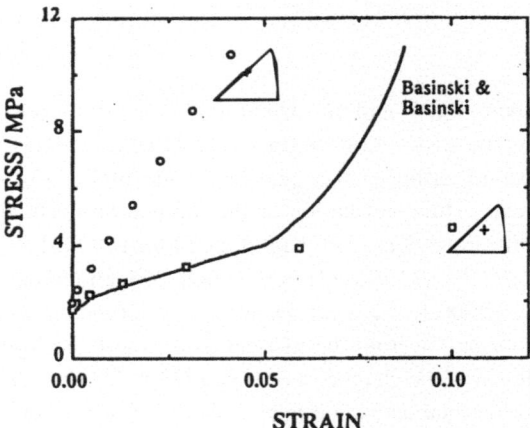

Fig. 3.9 Stress–strain relationships for Cu single crystals obtained when the dislocation mean free path is given by Eq. (3.23). $c^* = 100$ and $\rho_0 = 1 \times 10^9$ m^{-2}. Diagonal components of the weight matrix **w** were set to 0 and off-diagonal components were chosen to satisfy the conditions of quasi-isotropy. Experimental data of Basinski and Basinski [3] are shown in solid line [10]

results obtained in the analysis with the model of Seeger et al., fine activation of secondary slip systems was observed and the mean free path of dislocations on the primary system was reduced, but the change was weak and gradual. The model given in Eq. (3.23) is still needed to elaborate further.

3.2 Geometrically Necessary Dislocations

Increment of the statistically stored dislocations (SS dislocations) was calculated from the increment of plastic shear strain, as seen in the previous section. The quantity evaluated in this way was a scalar density in each slip system; no character or direction of dislocation lines were determined. The sum of Burgers vector of SS dislocations as a whole was zero. In contrast, the density of the geometrically necessary dislocations (GN dislocations) [16] is not evaluated directly from the plastic shear strain but is derived from its spatial gradient. Densities for the edge and screw components are obtained individually and therefore not only the density but also the direction of dislocation line is determined.

Let us review the basic concept of GN dislocations. Suppose that there are regions 1–4 continuously in the crystal, as shown in Fig. 3.10a. The width and height are $\Delta \xi$ and $\Delta \eta$, respectively. Edge dislocations have passed through regions 1 and 2 and exist in the region 3 but have not yet entered region 4. If the plastic shear strain due to the passage of the dislocations is $\Delta \gamma$ as shown in Fig. 3.10b, there is a gradient in plastic shear strain in region 3 because the strain is zero in region 4. Dislocations "must" be in region 3.

Let us find the density of dislocations in region 3. The average gradient of plastic shear strain in region 3 is noted as $\partial \gamma / \partial \xi$. Multiplying this gradient by $\Delta \xi$, the product is the difference in plastic shear strain between the left and right edges of region 3. Multiplying further by the height $\Delta \eta$, then the product $-\frac{\partial \gamma}{\partial \xi} \Delta \xi \Delta \eta$ is equal to the plastic displacement

3.2 Geometrically Necessary Dislocations

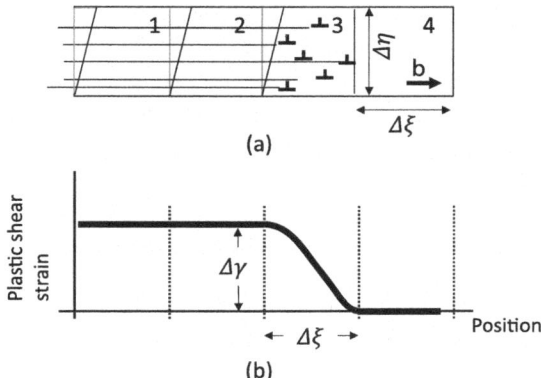

Fig. 3.10 Plastic slip deformation and geometrically necessary dislocations. **a** Assume that there are regions 1–4 and dislocations pass through the regions 1 and 2 and exist in 3. Dislocations have not yet entered the region 4, **b** distribution of the plastic shear strain is a constant in the regions 1 and 2, while the strain is zero in the region 4. The strain varies in the region 3

produced at the left upper corner of region 3, as shown in Fig. 3.11a. While, if the density of dislocations in region 3 is given by ρ_G, the amount of discontinuity detected by the Burgers circuit drawn along the periphery of region 3 is $\rho_G \Delta \xi \Delta \eta \times b$, because the number of dislocations in region 3 is $\rho_G \Delta \xi \Delta \eta$. This quantity must be equal to the difference in plastic displacement shown in Fig. 3.11a, leading to the following relationship between dislocation density and the spatial gradient of plastic shear strain [16],

$$\rho_G = -\frac{1}{b}\frac{\partial \gamma}{\partial \xi}. \tag{3.25}$$

There are a rich amount of researches on dislocation density tensor by Kondo, Nye, Kroener and Mura [17], and the GN dislocations has a close relation to the dislocation

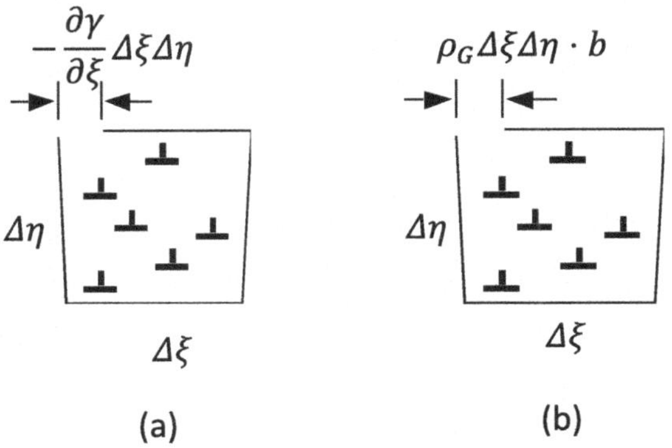

Fig. 3.11 a If the spatial gradient of the shear strain is $\frac{\partial \gamma}{\partial \xi}$ in the region 3 shown in Fig. 3.10, difference of plastic displacement between the left and right upper corners is $\frac{\partial \gamma}{\partial \xi}\Delta \xi \Delta \eta$, **b** when the density of dislocations in the region 3 is ρ_G, the discontinuity depicted by the Burgers' circuit drawn on the periphery of the region 3 is $\rho_G \Delta \xi \Delta \eta \cdot b$

density tensor. The theory of dislocation density tensor has a general form where crystal structure, slip plane, or slip direction are not specified but spatial gradient of plastic distortion is connected to the dislocation density tensor. The GN dislocations are considered to be the quantity that maps the dislocation density tensor to a coordinate system defined by the slip plane and slip direction. Equation (3.25) shows that the density of edge dislocations is proportional to the spatial gradient of plastic shear strain in the direction of the Burgers vector. Mapping the dislocation density tensor to the coordinate system defined by slip plane and slip direction, we obtain another component of screw dislocations [18],

$$\rho_{G,edge}^{(n)} = -\frac{1}{b}\frac{\partial \gamma^{(n)}}{\partial \xi^{(n)}}, \qquad (3.26a)$$

$$\rho_{G,screw}^{(n)} = \frac{1}{b}\frac{\partial \gamma^{(n)}}{\partial \zeta^{(n)}}. \qquad (3.26b)$$

Equations (3.26a) and (3.25) are identical, but sub- and superscripts are added to indicate the character of the component and the slip system to which they belong. Equation (3.26b) defines the density of the screw component. ζ is the direction perpendicular to the Burgers vector on the slip plane. Figure 3.12a–c schematically illustrate Eq. (3.26a). In Fig. 3.12a, we look down the slip plane with rectangular dislocation loops a–b–c–d and assume that plastic shear strain is realized by the expansion of these loops. The directions parallel and perpendicular to the Burgers vector are ξ and ζ, respectively. Figure 3.12b and c show the profiles of plastic shear strain observed along lines AA' and BB' parallel to ζ and ξ, respectively. The plastic shear strain is constant inside the dislocation loops and zero outside, and GN dislocations of the density given by Eq. (3.26b) must exist along the dislocation loops.

GN dislocations are not necessarily pure edge or pure screw ones. The density of mixed character is given by the norm,

$$\left\| \rho_G^{(n)} \right\| = \sqrt{\left(\rho_{G,edge}^{(n)} \right)^2 + \left(\rho_{G,screw}^{(n)} \right)^2}. \qquad (3.27)$$

The tangent vector of dislocation line **l**, and characteristic angle φ between the direction of dislocation line and the Burgers vector, shown in Fig. 3.13 are given as follows [19, 20],

$$\mathbf{l}^{(n)} = \frac{1}{\left\| \rho_G^{(n)} \right\|} \left(\rho_{G,screw}^{(n)} \mathbf{b}^{(n)} + \rho_{G,edge}^{(n)} \mathbf{b}^{(n)} \times \mathbf{n}^{(n)} \right), \qquad (3.28)$$

$$\varphi = arctan \frac{\rho_{G,edge}^{(n)}}{\rho_{G,screw}^{(n)}}. \qquad (3.29)$$

3.2 Geometrically Necessary Dislocations

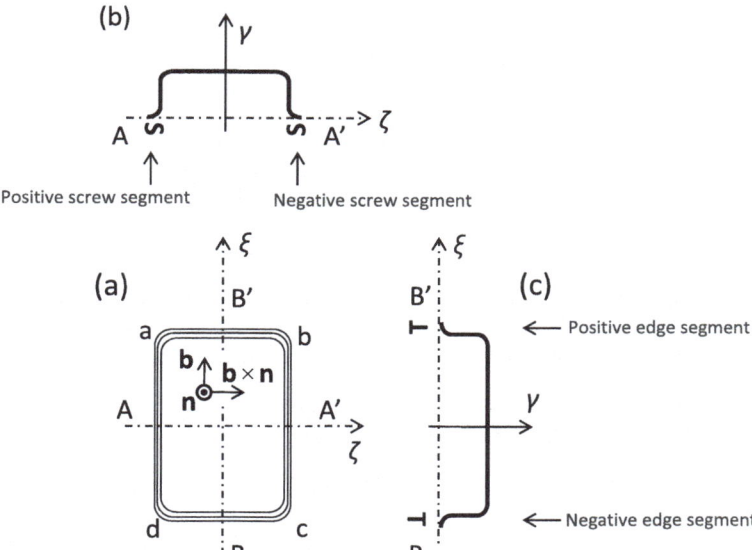

Fig. 3.12 **a** Rectangular shaped dislocation loops **a–b–c–d** are assumed to be formed on the slip plane. Plastic slip took place inside the loops. **b** and **c** profiles of plastic shear strain along lines AA' and BB' will be something like those shown in the figures with flat tops. Positive and negative dislocations of screw and edge characters must exist where the value of plastic shear strain change

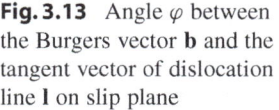

Fig. 3.13 Angle φ between the Burgers vector **b** and the tangent vector of dislocation line **l** on slip plane

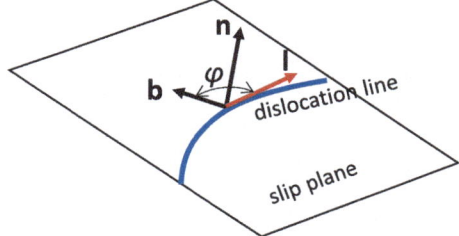

Here, $\mathbf{b}^{(n)}$ and $\mathbf{n}^{(n)}$ denote the unit vectors parallel to the Burgers vector and normal to the slip plane, respectively.

Equations (3.27), (3.28) (3.29) are useful for visualizing GN dislocations. Figure 3.14 shows an example [21]. As shown in Fig. 3.14a, the specimen is a cube-shaped single crystal with lateral dimension 30 μm and a spherical void region of 4 μm in diameter is introduced. Uniform tensile displacements were applied to six surfaces of the specimen to generate a hydrostatic stress field. Stress concentrates near the void, resulting in a localized slip deformation and the formation of GN dislocations.

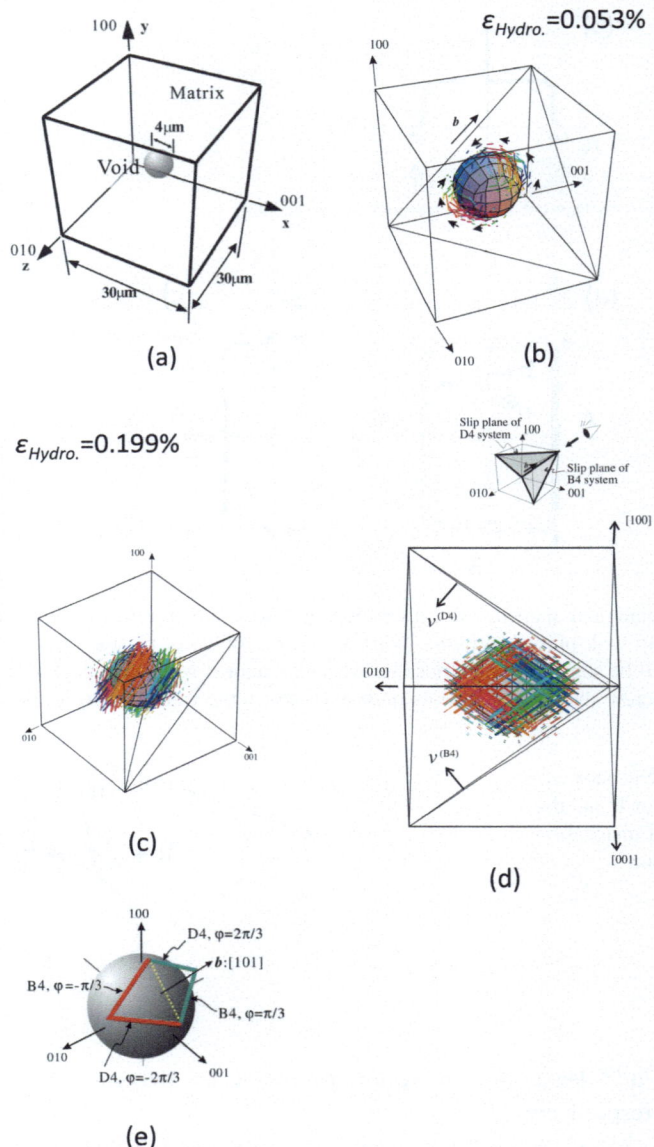

Fig. 3.14 An example of visualization of geometrically necessary dislocations. **a** A sphere shaped void area with a diameter of 4 μm is placed at the centre of the cube shaped matrix region. **b** and **c** GN dislocations formed in B4 slip system when the hydrostatic strain is 0.053% and 0.199%, respectively, **d** superimposed view of GN dislocations in B4 and D4 slip systems, where the two slip systems share the slip direction and cross-slip systems to each other, **e** schematic illustration of the GN dislocations in B4 and D4 slip systems [21]

Figure 3.14b shows distribution of GN dislocations on B4: $(11\bar{1})[101]$ slip system when the nominal volumetric strain is 0.053%. A short line segment is drawn at the center of each finite element. The direction of the line segment is determined by Eq. (3.28) and its length and thickness are determined by the product of the density given by Eq. (3.27) and the element volume. Line segments are assigned a color as a function of their characteristic angle. It can be seen that half shear loops are generated from the void region. Figure 3.14c shows the GN dislocations on slip systems B4: $(11\bar{1})[101]$ at a nominal strain of 0.199%. Figure 3.14d shows GN dislocations on B4 and D4: $(\bar{1}11)[101]$ slip systems superimposed. These slip systems are in a cross-slip relationship with each other. Analysis results showed that the characteristic angles of GN dislocations on B4 and D4 slip systems were $\pm\pi/3$ and $\pm 2\pi/3$, respectively. The extra half planes of the GN dislocations on these slip systems were found to be positioned inside the loop structure and the loop enclose extra half planes. Figure 3.14e schematically shows the character of line segments of GN dislocations and the direction of Burgers vector. The loops are analogous to prismatic dislocation loops [12] except that the loops are not on flat planes but form a shape of bended rhombus. Formation of the structure similar to this occur in other combinations of two slip systems which are in the relationship of cross-slip to each other and they mimic the emission of prismatic dislocation loops.

Another interesting feature of GN dislocations is that the density is scale-dependent because the density is obtained by the spatial gradient of plastic shear strain. Even if the amount of change in the shear strain $\Delta\gamma$ is the same in the region 3 of Fig. 3.10, different density of GN dislocations is obtained if the distance $\Delta\xi$ is not the same. This fact is important for understanding the experimental fact that mechanical properties depend on the size of microstructure, such as average grain size. This point is discussed in Sect. 4.2.1.

References

1. Kocks UF, Mecking H (2003) Prog Mater Sci 48:171
2. Seeger A, Diehl J, Mader S, Rebstock H (1957) Philos Mag A J Theor Exp Appl Phys 2:323
3. Basinski ZS, Basinski SJ (1964) Philos Mag A J Theor Exp Appl Phys 9:51
4. Young FW (1962) J Appl Phys 33:963
5. Fourie JT (1968) Philos Mag A J Theor Exp Appl Phys 17:735
6. Van Bueren VG (1960) Imperfections in Crystals. NorthHolland, Amsterdam
7. Ohashi T (1987) J Japan InstMetals 51:37
8. Ohashi T (1987) Trans JIM 28:906
9. Ohashi T (1994) Phil Mag A 70:793
10. Ohashi T. Numerical Modeling of Plastic Multi Slip in F.C.C. Crystals, in:. Tokuda M (Ed.). Proc. I MMM' 9 3 Int. Semin. Microstruct. Mech. Prop. New Eng. Mater. Mie: Mie Academic Press; 1993.
11. Higashida K, Takamura J, Narita N (1986) Mater Sci Eng 81:239
12. Cottrell AH (1952) Dislocations and Plastic Flow in Crystals, 1st editio. Oxford University Press, Oxford

13. Diehl J (1956) Int J Mater Res 47:331
14. Kuhlmann-Wilsdorf D (1989) Mater Sci Eng A. 113:1
15. Kujirai S, Shizawa K (2020) Philos Mag 100:2106
16. Ashby MF (1970) Philos Mag 21:399
17. Mura T (1987) Micromechanics of Defects in Solids, Second, re. Kluwer Academic Press, Dordrecht
18. Ohashi T (1997) Philos Mag Lett 75:51
19. Ohashi T (1999) J Phys IV Fr; 9:Pr9-279.
20. Ohashi T (2004) Int J Plast 20:1093
21. Ohashi T (2005) Int J Plast 21:2071

Models for the Critical Resolved Shear Stress

Abstract

Taylor model for the critical resolved shear stress which is proportional to the square root of the accumulated dislocation density is reviewed from one-dimensional perspective and extended to three-dimensional space. Interactions between slip systems, as well as slip and twin systems are formally summarized into the interaction matrix between deformation modes. Interaction matrices for FCC, BCC, and HCP crystals are considered. Interaction between twin and slip systems in HCP crystals is reviewed with reference to the existing literature. The effect of length scale of metal microstructure on strength and strain hardening is first introduced from the scale dependent nature of geometrically necessary dislocations and then detailed process of dislocation motion inside grains of finite size is examined by dislocation dynamics simulation. The results are incorporated into the model of critical resolved shear stress. Effect of grain boundaries which impede movement of dislocations is introduced also to the model of dislocation mean free path.

Keywords

Critical resolved shear stress · Length scale of metal microstructure · Taylor model · Scale dependent strain hardening · Scale dependent yield stress · Orowan stress in a finite medium · Slip and twinning · Interaction matrix · FCC crystals · BCC crystals · HCP crystals

4.1 Dislocation Accumulation and Strain Hardening

4.1.1 One-Dimensional Approach

In a monograph [1] on the plastic slip and dislocations, Cottrell discusses how the movement of dislocations are obstructed by accumulated ones and how the inhomogeneity of slip deformation and activity of secondary slip systems affect the slip deformation. An attempt was made to model the effect of fine activity of secondary system in Sect. 3.1.2.2. In the present section, let us further consider the modeling of the suppression of moving dislocations by accumulated dislocations. The suppression effect of accumulated dislocations to the moving dislocation is mediated by the stress field or comes from the energy needed to cut dislocations each other.

First, let us consider the effect of the stress field. The stress field here is an internal stress field formed in the microstructure by accumulated dislocations, etc. If it is integrated over the entire microstructure, it becomes zero, but locally it is distributed in the microstructure in such a way that it inhibits, accelerates, or have no effect on the motion of dislocations. As the simplest case, an edge dislocation moves near another edge dislocation belonging to the same slip system (Fig. 4.1). The dislocation is accompanied by a self-stress field, and the combined stress field formed when the two dislocations are close to each other depends on their arrangement. When an external stress is applied, the arrangement of the two dislocations changes so that the total strain energy is minimized. However, when the external stress exceeds a threshold, the two dislocations are no longer stable under the stress and they pass each other.

If the distance between the slip planes is l as shown in Fig. 4.1, the stress required for the dislocation to pass is,

$$\tau = \mu b / 8\pi (1 - \nu) l. \tag{4.1}$$

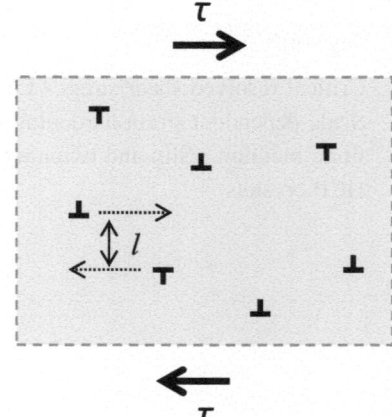

Fig. 4.1 Interaction of dislocations moving on parallel slip planes via elastic stress field

4.1 Dislocation Accumulation and Strain Hardening

On the other hand, if the density of accumulated dislocations is ρ, the average spacing is

$$l = 1/\sqrt{\rho}. \tag{4.2}$$

Therefore, Eq. (4.1) is equivalent to the Taylor model,

$$\theta = \alpha\mu b \sqrt{\rho}. \tag{4.3}$$

As is well known, the size of the region affected by the self-stress field of the dislocation is considerably larger than the length of the Burgers vector. Therefore, the contribution from the thermal vibration of the atoms to the elastic interaction of dislocations is negligible.

For the case where a moving dislocation cut accumulated dislocations (Fig. 4.2), the sum of the formation energy of a jog of one Burgers vector length formed at the dislocation and the work done by the dislocation is,

$$U(\tau) = \alpha\mu b^3 - \tau l b^2. \tag{4.4}$$

From the condition that this energy is zero, the stress required for the cutting to occur is,

$$\theta = \alpha\mu b/l. \tag{4.5}$$

Fig. 4.2 Schematic illustration of a moving dislocation intersecting accumulated ones

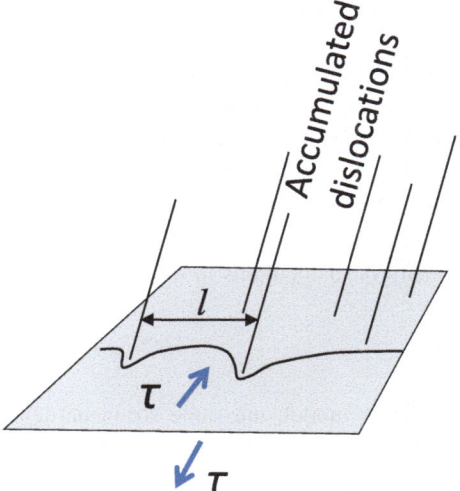

Here, l is the distance between forest dislocations, as shown in Fig. 4.2, and the stress required to cut forest dislocations is proportional to the square root of the accumulated density, and this is given by Eq. (4.3).

The deformation resistance due to this cutting process is considered to be dependent on temperature [1]. This differs from the case of elastic interactions via the self-stress field of dislocations. A model is proposed that gives the dependence of the Taylor term on temperature and strain rate. This will be briefly discussed in Sect. 4.1.3 in conjunction with the interaction matrix of slip systems.

Accumulated dislocations are not the only factors that inhibit dislocation motion. In crystal plasticity analysis, the Peierls stress on dislocation motion, drag resistance due to solute atoms, and other deformation resistances that do not change in magnitude as deformation proceeds are collectively called lattice friction stresses and are added to the deformation resistance caused by the accumulated dislocations,

$$\theta = \theta_0(T) + a\mu b\sqrt{\rho_S}, \tag{4.6}$$

where, the first term on the right-hand side is the lattice friction stress. Since a thermal activation process is often involved when dislocations overcome the physical reality that causes lattice frictional stress, it is explicitly written as $\theta_0(T)$.

4.1.2 Extension to Three-Dimensional Space and Inclusion of the Effect of Twin Deformation

How can the critical resolve shear stress be evaluated when various slip systems, and even twin systems, may be active simultaneously or overlapping? As Cottrell [1] points out, hardening of a slip system is not merely the result of slip process taking place in that system but the result of slip process in its whole neighborhood. Let us first consider the multiplication of slips. One simple model would be to take the sum of the density of dislocations that accumulate in the various slip systems and use it as the representative density of dislocations in the Taylor term.

$$\theta^{(n)} = \theta_0(T) + a\mu b \sqrt{\sum_{m=1}^{N} \rho_S^{(m)}}. \tag{4.7}$$

In this model, the same strain hardening occurs in all slip system whether they are active or not and the strain hardening is isotropic. If we introduce a matrix $\chi^{(nm)}$ that gives the magnitude of contribution of dislocations accumulated in the slip system m,

$$\theta^{(n)} = \theta_0(T) + a\mu b \sqrt{\sum_{m=1}^{N} \chi^{(nm)} \rho_S^{(m)}}, \tag{4.8}$$

4.1 Dislocation Accumulation and Strain Hardening

then, the strain hardening of slip system n depends on the density of dislocations accumulated in slip system m and the intensity of their contribution, resulting in an anisotropic hardening.

Another model to give different contribution from slip systems has the form,

$$\theta^{(n)} = \theta_0^{(n)}(T) + \sum_{m=1}^{N} \Omega^{(nm)} a\mu b^{(m)} \sqrt{\rho_S^{(m)}}, \qquad (4.9)$$

where, contributions from each slip systems are individually given by Taylor terms and the sum of the contributions gives the CRSS. The interaction matrix $\Omega^{(nm)}$ defines the magnitude of contribution from the Taylor term of slip system m to the CRSS of the slip system n.

4.1.3 Deformation Modes in FCC, BCC and HCP Crystals and Their Interaction Matrix

Table 4.1 summarizes the slip and twin systems often observed in FCC (face-centered cubic), BCC (body-centered cubic) and HCP (hexagonal close packed) crystals. In this section, we will first review the slip interaction matrices for FCC and BCC crystals, and then discuss some possibilities for the case of crystals with HCP structure regarding slip and its interaction, deformation resistance of twin deformations, and their interactions.

In FCC crystals, slip mostly occurs on {111} slip plane and in <110> direction. There are four slip planes of (111), ($\bar{1}$11), (11$\bar{1}$), (1$\bar{1}$1) and three slip directions are possible on each slip plane, resulting in twelve possible slip systems. Table 4.2 shows slip planes and directions of the twelve slip systems and the notation by Schmid and Boas. Figure 4.3a shows Thompson tetrahedron where four faces consist of {111} crystal planes and three edges bounding each surface are parallel to the slip direction. As already mentioned, interactions between self-systems and coplanar systems is mediated by elastic stress field and dependence to temperature is small. The interaction with other slip systems involves mutual cutting of dislocation lines and the dependence to temperature and strain rate will not be negligible. Following the study by Kocks, Mecking and Estrin [3–6], such dependence could be modeled by writing the component of the interaction matrix as,

$$\Omega^{(ij)} = \Omega_0^{(ij)} \left(\frac{\dot{\gamma}}{\dot{\gamma}_0} \right)^p. \qquad (4.10)$$

Here, $\Omega_0^{(ij)}$ and $\dot{\gamma}_0$ are reference values for the component of interaction and reference strain rate, while p is the exponent of the strain rate dependency. The strain rate is a function of temperature and stress. The strain rate-independent property of the interaction with the self and with the coplanar systems will be given by $p = 0$.

Table 4.1 Shear planes and shear directions and twin strain of FCC, BCC, and HCP crystals. Data for HCP crystals are taken from Narita [2] and partially modified. c and a are the lattice constants of the crystal axes in the **c** and **a** directions, respectively

Crystal structure	Deformation mode	Shear plane	Shear direction	Crystallographically determined twin strain, Γ_{twin}	Materials
FCC	Slip	$\{11\bar{1}\}$	$\langle 101 \rangle$	–	
	Twin	$\{11\bar{1}\}$	$\langle 112 \rangle$	$1/\sqrt{2}$	Cu, Ag, Au, α-brass
BCC	Slip	$\{110\}$	$\langle \bar{1}11 \rangle$	–	
		$\{1\bar{1}2\}$	$\langle \bar{1}11 \rangle$	–	
	Twin	$\{2\bar{1}1\}$	$\langle 11\bar{1} \rangle$	$1/\sqrt{2}$	α-Fe, V, Nb, Ta, Mo-35%Re
HCP	Slip	$\{0001\}$	$\langle 11\bar{2}0 \rangle$	–	
	Slip	$\{1\bar{1}00\}$	$\langle 11\bar{2}0 \rangle$	–	
	Slip	$\{1\bar{1}01\}$	$\langle 11\bar{2}0 \rangle$	–	
	Slip	$\{1\bar{1}01\}$	$\langle \bar{1}2\bar{1}3 \rangle$	–	
	Slip	$\{11\bar{2}2\}$	$\langle \bar{1}\bar{1}23 \rangle$	–	
	Twin	$\{10\bar{1}2\}$	$\langle 10\bar{1}\bar{1} \rangle$	$\dfrac{(c/a)^2 - 3}{\sqrt{3}c/a}$	Cd, Zn, Mg, Co, Be, Zr, Ti
	Twin	$\{10\bar{1}1\}$	$\langle \bar{1}012 \rangle$	$\sqrt{3}a/c$	Mg
	Twin	$\{11\bar{2}2\}$	$\langle 11\bar{2}\bar{3} \rangle$	$(2a/3c)\{(c/a)^2 - 2\}$	Zr, Ti

Table 4.2 Twelve slip systems in FCC crystals

Slip system number	Notation	Slip plane	Slip direction
1	A2	(111)	$[1\bar{1}0]$
2	A6	(111)	$[01\bar{1}]$
3	A3	(111)	$[10\bar{1}]$
4	D1	$(\bar{1}11)$	$[110]$
5	D6	$(\bar{1}11)$	$[01\bar{1}]$
6	D4	$(\bar{1}11)$	$[101]$
7	B2	$(11\bar{1})$	$[1\bar{1}0]$
8	B5	$(11\bar{1})$	$[011]$
9	B4	$(11\bar{1})$	$[101]$
10	C1	$(1\bar{1}1)$	$[110]$
11	C5	$(1\bar{1}1)$	$[011]$
12	C3	$(1\bar{1}1)$	$[10\bar{1}]$

4.1 Dislocation Accumulation and Strain Hardening

Fig. 4.3 **a** Thompson's tetrahedra showing 4 slip planes and 3 slip directions on each plane in FCC crystal lattice, **b** dislocation interaction between different slip systems depicted on the Thompson's tetrahedra

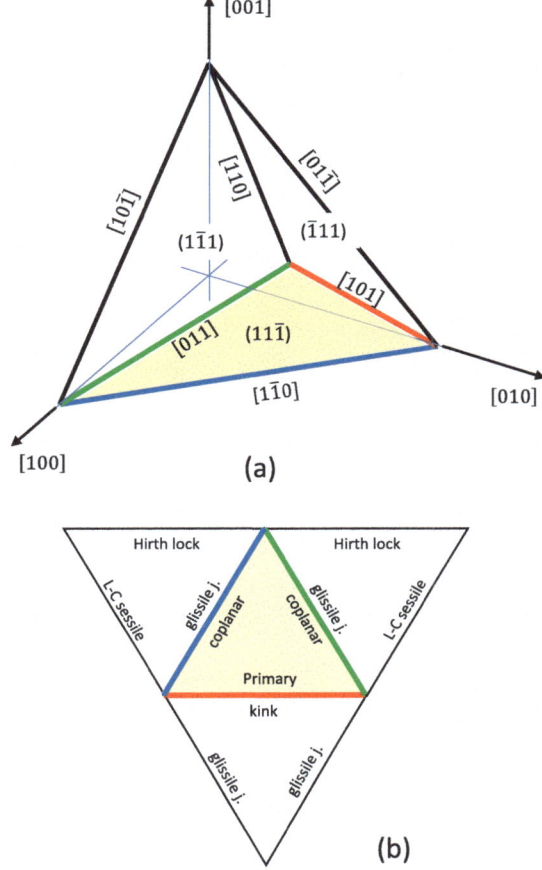

Among combinations of slip systems, slip systems that share a common slip direction can cross-slip each other. Taking these considerations into account, a 12×12 interaction matrix could be drawn as shown in Fig. 4.4, where n is the active slip system and m is the slip system that interacts with it. How can we quantify the component values of this interaction matrix?

The interaction between dislocations in crystals of FCC structure has been studied [7]. The case of a dislocation resolved into two Shockley partial dislocations on the {111} plane reacting to form a strong immobile dislocation is known as an important one. Other dislocation interactions belonging to 12 slip systems are also known, which are classified into six types as shown in Table 4.3. The first two are elastic interactions and the remaining four are interactions when dislocations cut each other. In Fig. 4.3a, assuming that the primary slip system is $(11\bar{1})[101]$, the slip plane on which the dislocation moves is depicted in light brown and the slip direction is in red. The interactions with dislocations

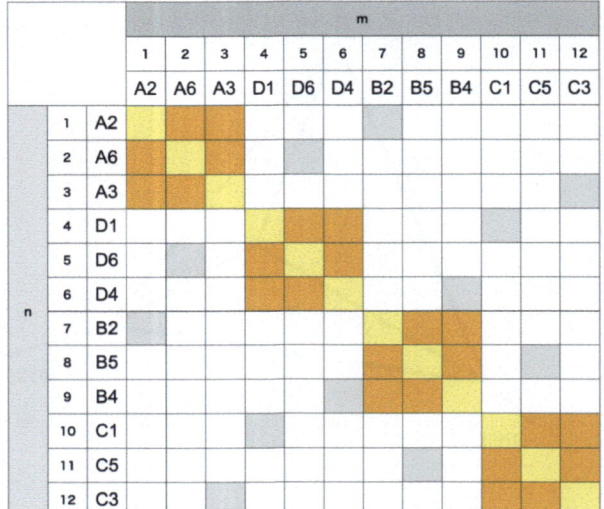

Fig. 4.4 Interaction matrix of 12 slip systems of {111}<110> in FCC crystals

Table 4.3 Interaction between dislocations on {111}<110> slip systems in FCC crystals

Dislocation interaction		$\Omega^{(nm)}$
Elastic interaction	Primary as against primary	r_0
↑	Primary as against coplanar	r_1
Junction formation	Hirth lock formation	r_2
↑	Glissile jog	r_3
↑	Lomer-Cottrell sessile dislocations	r_{31}
↑	Kink formation	r_4

belonging to the 12 slip systems are shown in Fig. 4.3b, which is a plane-width expansion of the Thompson tetrahedron. The same procedure can be repeated if the dislocations of interest belong to a different slip system. In other words, the interaction with other dislocations can be determined systematically by placing the slip system of interest at the position of "Primary" in Fig. 4.3b. In this way, all combinations of 12×12 interactions can be classified into six types of interactions.

The strength of slip resistance generated by the six interactions is represented by r_0, r_1, r_2, r_3, r_{31}, and r_4, as shown in Table 4.3. Based on Fig. 4.3 and the geometric arrangement among the slip systems described above, we can determine the interaction

4.1 Dislocation Accumulation and Strain Hardening

			m											
			1	2	3	4	5	6	7	8	9	10	11	12
			A2	A6	A3	D1	D6	D4	B2	B5	B4	C1	C5	C3
n	1	A2	r_0	r_1	r_1	r_2	r_3	r_{31}	r_4	r_3	r_3	r_2	r_{31}	r_3
	2	A6	r_1	r_0	r_1	r_3	r_4	r_3	r_3	r_2	r_{31}	r_{31}	r_2	r_3
	3	A3	r_1	r_1	r_0	r_{31}	r_3	r_2	r_3	r_{31}	r_2	r_3	r_3	r_4
	4	D1	r_2	r_3	r_{31}	r_0	r_1	r_1	r_2	r_{31}	r_3	r_4	r_3	r_3
	5	D6	r_3	r_4	r_3	r_1	r_0	r_1	r_{31}	r_2	r_3	r_3	r_2	r_{31}
	6	D4	r_{31}	r_3	r_2	r_1	r_1	r_0	r_3	r_3	r_4	r_3	r_{31}	r_2
	7	B2	r_4	r_3	r_3	r_2	r_{31}	r_3	r_0	r_1	r_1	r_2	r_3	r_{31}
	8	B5	r_3	r_2	r_{31}	r_{31}	r_2	r_3	r_1	r_0	r_1	r_3	r_4	r_3
	9	B4	r_3	r_{31}	r_2	r_3	r_3	r_4	r_1	r_1	r_0	r_{31}	r_3	r_2
	10	C1	r_2	r_{31}	r_3	r_4	r_3	r_3	r_2	r_3	r_{31}	r_0	r_1	r_1
	11	C5	r_{31}	r_2	r_3	r_3	r_2	r_{31}	r_3	r_4	r_3	r_1	r_0	r_1
	12	C3	r_3	r_3	r_4	r_3	r_{31}	r_2	r_{31}	r_3	r_2	r_1	r_1	r_0

Fig. 4.5 Interaction matrix of twelve slip systems in FCC crystals filled with matrix component

when each of the 12 slip systems is the primary one. Expanding this into an interaction matrix yields Fig. 4.5 [8, 9].

Although the specific values of the matrix components r_0, r_1, r_2, r_3, r_{31}, and r_4 have not yet been fully elucidated, the following relationships are inferred from experimental results [10] on latent hardening of copper single crystals [9, 11],

$$r_0 \leq r_1 \leq r_2 \leq r_3 = r_{31} = r_4, \tag{4.11}$$

or

$$r_0 \leq r_1 = r_2 = r_4 \leq r_3 \leq r_{31}. \tag{4.12}$$

In BCC crystals, the slip planes are often {110} or {112} and the slip directions are <111>. Figure 4.6a shows slip systems (110)[$\bar{1}$11] and (1$\bar{1}$2)[$\bar{1}$11]. The number of slip systems is 12 for both {110}<111> and {112}<111> systems for a total of 24. The combinations of slip planes and directions are shown in Table 4.4.

The interaction matrix of the 24 slip systems is shown in Fig. 4.6b, and it can be seen that there are only eight combinations of slip systems in a coplanar relationship, while there are 120 combinations in a cross-slip relationship.

Structure of HCP crystal lattice is schematically shown in Fig. 4.7a. Four-index Miller-Bravais indices are used to show orientation and plane in HCP crystals. Figure 4.7a shows

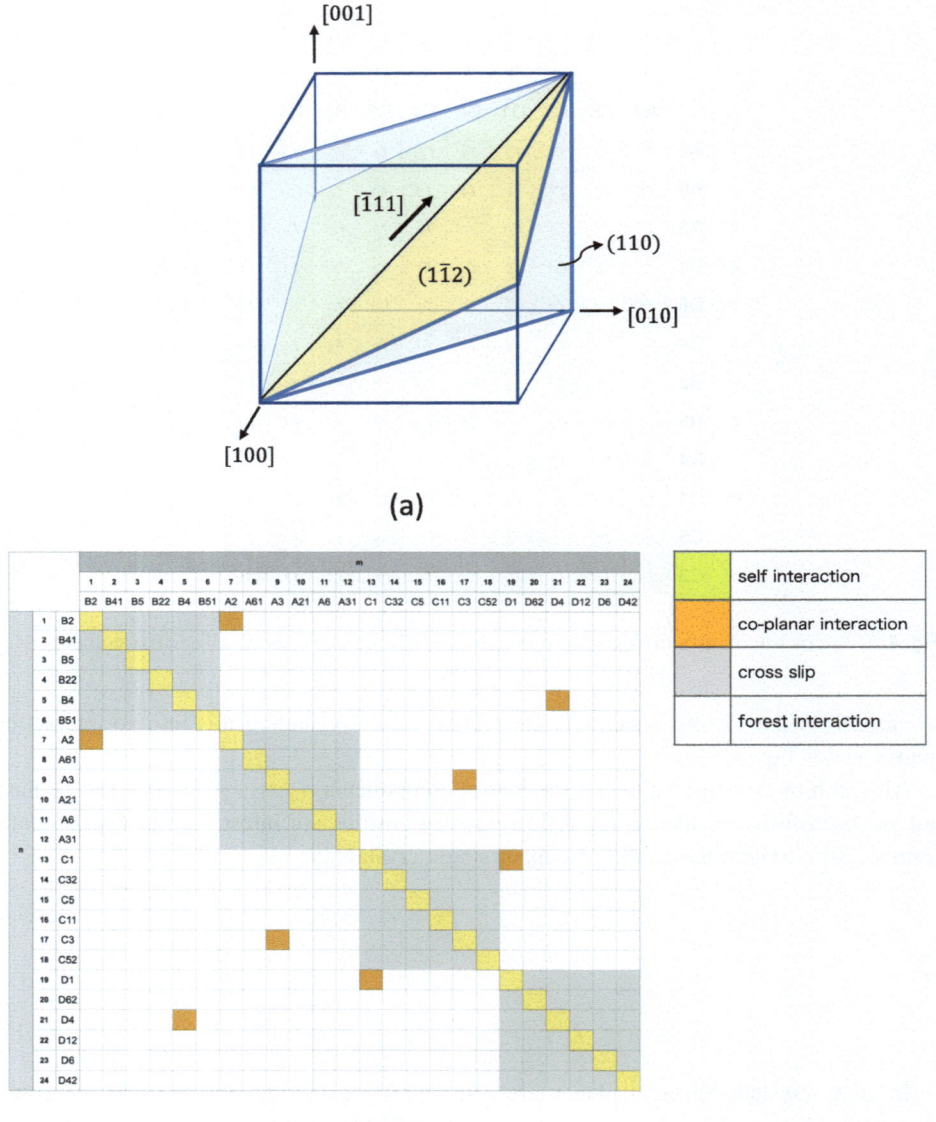

Fig. 4.6 **a** Slip systems (110) [$\bar{1}$11] and (1$\bar{1}$2)[$\bar{1}$11] in BCC crystals, **b** interaction matrix of 24 slip systems. Definition of slip systems is given in Table 4.4

4.1 Dislocation Accumulation and Strain Hardening

Table 4.4 Twenty four slip systems in BCC crystals

Slip system number	Notation	Slip plane	Slip direction
1	B2	$(0\bar{1}1)$	$[111]$
2	B41	$(\bar{1}2\bar{1})$	$[111]$
3	B5	$(\bar{1}10)$	$[111]$
4	B22	$(2\bar{1}\bar{1})$	$[111]$
5	B4	$(\bar{1}01)$	$[111]$
6	B51	$(\bar{1}\bar{1}2)$	$[111]$
7	A2	$(0\bar{1}1)$	$[\bar{1}11]$
8	A61	$(1\bar{1}2)$	$[\bar{1}11]$
9	A3	(101)	$[\bar{1}11]$
10	A21	(211)	$[\bar{1}11]$
11	A6	(110)	$[\bar{1}11]$
12	A31	$(12\bar{1})$	$[\bar{1}11]$
13	C1	(011)	$[11\bar{1}]$
14	C32	$(\bar{1}21)$	$[11\bar{1}]$
15	C5	$(\bar{1}10)$	$[11\bar{1}]$
16	C11	$(2\bar{1}1)$	$[11\bar{1}]$
17	C3	(101)	$[11\bar{1}]$
18	C52	(112)	$[11\bar{1}]$
19	D1	(011)	$[1\bar{1}1]$
20	D62	$(\bar{1}12)$	$[1\bar{1}1]$
21	D4	$(\bar{1}01)$	$[1\bar{1}1]$
22	D12	$(21\bar{1})$	$[1\bar{1}1]$
23	D6	(110)	$[1\bar{1}1]$
24	D42	(121)	$[1\bar{1}1]$

some basic directions and Miller-Bravais indices. Indices on basal plane are also shown in Fig. 4.7b. In addition to slip deformation, twinning activity also plays an important role in many cases in HCP crystals. Well-known slip and twin systems include the following.

- 3 basal <a> slip systems; $\{0001\}\langle 11\bar{2}0\rangle$
- 3 prismatic 1 <a> slip systems; $\{1\bar{1}00\}\langle 11\bar{2}0\rangle$
- 6 pyramidal 1 <a> slip systems; $\{1\bar{1}01\}\langle 11\bar{2}0\rangle$
- 12 pyramidal 1 <a + c> slip systems; $\{1\bar{1}01\}\langle \bar{1}2\bar{1}3\rangle$
- 6 pyramidal 2 <a + c> slip systems; $\{11\bar{2}2\}\langle \bar{1}\bar{1}23\rangle$
- 6 twin systems; $\{1\bar{1}02\}\langle \bar{1}101\rangle$

Fig. 4.7 **a** HCP lattice and Miller-Bravais indices, **b** Miller-Bravais indices on the basal plane

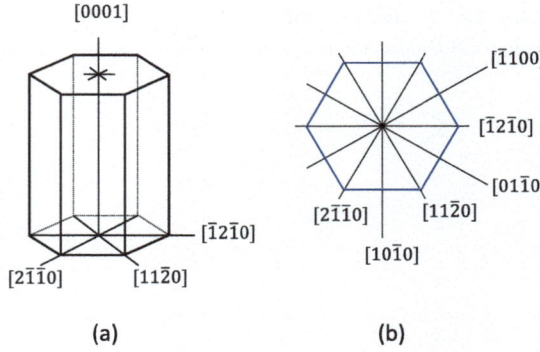

In total, there are 36 deformation modes in this case. Figure 4.8 shows one of each of these deformation modes. Figure 4.8a basal<a>slip; (0001)[$2\bar{1}\bar{1}0$] and prismatic 1<a>slip; ($01\bar{1}0$)[$2\bar{1}\bar{1}0$], (b) pyramidal 1<a>slip; ($10\bar{1}1$)[$\bar{1}2\bar{1}0$] and pyramidal 1<a + c>slip; ($10\bar{1}1$)[$\bar{2}113$], (c) pyramida 2<a + c>slip; ($2\bar{1}\bar{1}2$)[$\bar{2}113$], and (d) twin; ($10\bar{1}2$)[$\bar{1}011$].

Figure 4.9 formally shows the interaction between 36 deformation modes of HCP crystals. Active deformation mode is placed column wise and denoted by n and interacting deformation modes and their traces are denoted by m. When slip deformation mode is activated ($n = 1, \ldots 30$) and interact with other slip deformation modes, there are cases of self-, coplanar-, cross-slip-, as well as the forest-interactions. Coplanar interactions occur between basal slip systems and also between pyramidal 1<a> and pyramidal 1<a + c> slip systems.

Fig. 4.8 Typical deformation modes in HCP crystals, **a** basal<a>slip system; (0001)[$2\bar{1}\bar{1}0$] and prismatic<a>slip system; ($01\bar{1}0$)[$2\bar{1}\bar{1}0$], **b** pyramidal 1<a>slip system; ($10\bar{1}1$)[$\bar{1}2\bar{1}0$] and pyramidal 1<a + c>slip system; ($10\bar{1}1$)[$\bar{2}113$], **c** pyramidal 2<a + c>slip system; ($2\bar{1}\bar{1}2$)[$\bar{2}113$], and **d** twin system; ($10\bar{1}2$)[$10\bar{1}\bar{1}$]

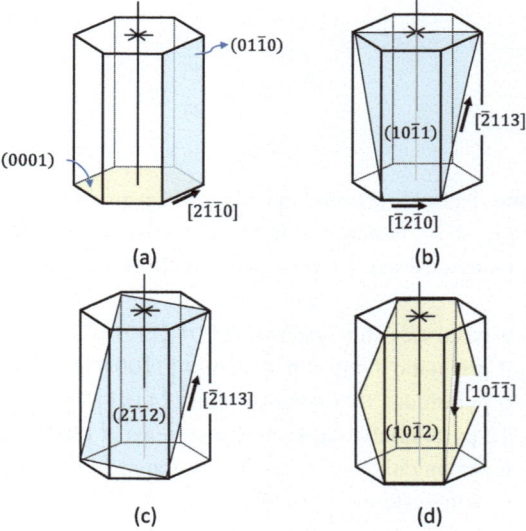

4.1 Dislocation Accumulation and Strain Hardening

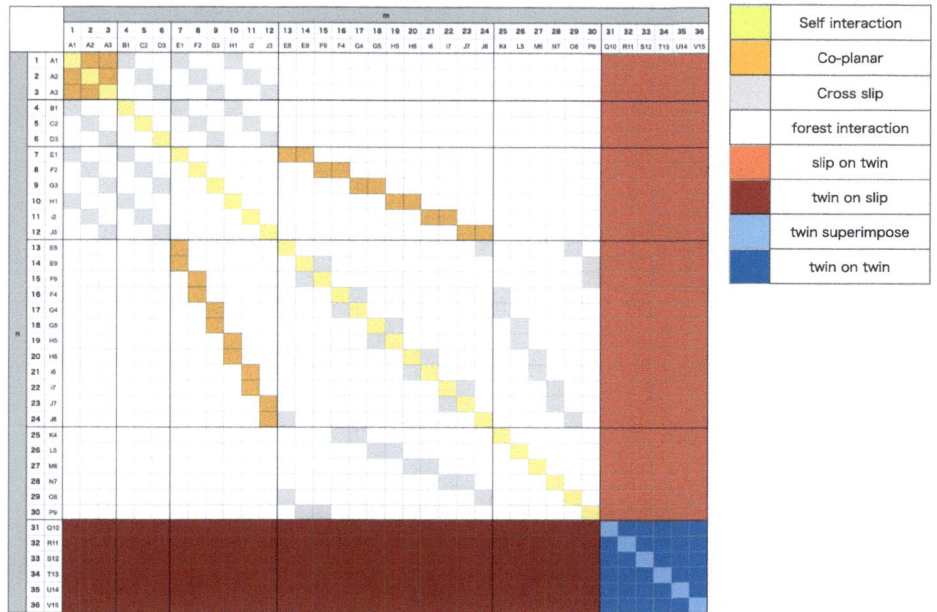

Fig. 4.9 Interaction matrix of 36 deformation modes in HCP crystals. Definitions of slip and twin systems are given in Table 4.5

When the cross-slip is discussed in a purely geometrical point of view, there are various combinations. First, the basal and prismatic systems shown in Fig. 4.8a share a slip direction and may cross-slip with each other. Also, pyramidal 1<a>, basal<a> and prismatic<a> share the slip direction $\langle 11\bar{2}0 \rangle$ and possible to cross slip. Pyramidal 1<a + c> shares slip direction with other pyramidal 1<a + c> and pyramidal 2<a + c> slip systems as shown in Fig. 4.10a and b and possibly cross-slip. On the other hand and away from the geometry of the slip system configuration, the probability of these cross slips depend on the expansion width of extended dislocations and other crystallographic factors and depend on materials, deformation temperature, strain rate etc. [12].

The interactions of slip ($n = 1,…30$) and twinning ($m = 31,…,36$) refer to the case when the slip deformation is introduced to the area where twinning deformation has already been introduced, usually resulting in a strong interaction of dislocations with twin boundaries. In the twinning deformation mode, $n = 31,…36$, there are cases where twins are formed in areas where slip dislocations have already accumulated, twinning superimposes in areas where other twin deformation mode has already took place, or continuation of twin deformation. Details of these process is closely connected to the CRSS of the deformation mode and strain hardening, on which we will discuss some more in Sect. 4.2.3.

Table 4.5 Slip and twin deformation modes in HCP crystals

Deformation mode number	Notation	Shear plane	Shear direction	Name for the deformation mode
1	A1	(0001)	$[11\bar{2}0]$	Basal <a> slip
2	A2	(0001)	$[\bar{1}2\bar{1}0]$	
3	A3	(0001)	$[\bar{2}110]$	
4	B1	$(1\bar{1}00)$	$[11\bar{2}0]$	Prism 1 <a> slip
5	C2	$(10\bar{1}0)$	$[\bar{1}2\bar{1}0]$	
6	D3	$(01\bar{1}0)$	$[\bar{2}110]$	
7	E1	$(1\bar{1}01)$	$[11\bar{2}0]$	Pyramidal 1<a> slip
8	F2	$(10\bar{1}1)$	$[\bar{1}2\bar{1}0]$	
9	G3	$(01\bar{1}1)$	$[\bar{2}110]$	
10	H1	$(\bar{1}101)$	$[11\bar{2}0]$	
11	I2	$(\bar{1}011)$	$[\bar{1}2\bar{1}0]$	
12	J3	$(0\bar{1}11)$	$[\bar{2}110]$	
13	E8	$(1\bar{1}01)$	$[\bar{1}2\bar{1}3]$	Pyramidal 1<a+c> slip
14	E9	$(1\bar{1}01)$	$[\bar{2}113]$	
15	F9	$(10\bar{1}1)$	$[\bar{2}113]$	
16	F4	$(10\bar{1}1)$	$[\bar{1}\bar{1}23]$	
17	G4	$(01\bar{1}1)$	$[\bar{1}\bar{1}23]$	
18	G5	$(01\bar{1}1)$	$[1\bar{2}13]$	
19	H5	$(\bar{1}101)$	$[1\bar{2}13]$	
20	H6	$(\bar{1}101)$	$[2\bar{1}\bar{1}3]$	
21	I6	$(\bar{1}011)$	$[2\bar{1}\bar{1}3]$	
22	I7	$(\bar{1}011)$	$[11\bar{2}3]$	
23	J7	$(0\bar{1}11)$	$[11\bar{2}3]$	
24	J8	$(0\bar{1}11)$	$[\bar{1}2\bar{1}3]$	
25	K4	$(11\bar{2}2)$	$[\bar{1}\bar{1}23]$	Pyramidal 2 <a+c> slip
26	L5	$(\bar{1}2\bar{1}2)$	$[1\bar{2}13]$	
27	M6	$(\bar{2}112)$	$[2\bar{1}\bar{1}3]$	
28	N7	$(\bar{1}\bar{1}22)$	$[11\bar{2}3]$	
29	O8	$(1\bar{2}12)$	$[\bar{1}2\bar{1}3]$	
30	P9	$(2\bar{1}\bar{1}2)$	$[\bar{2}113]$	

(continued)

4.1 Dislocation Accumulation and Strain Hardening

Table 4.5 (continued)

Deformation mode number	Notation	Shear plane	Shear direction	Name for the deformation mode
31	Q10	$(1\bar{1}02)$	$[\bar{1}101]$	Twin deformation
32	R11	$(10\bar{1}2)$	$[\bar{1}011]$	
33	S12	$(01\bar{1}2)$	$[0\bar{1}11]$	
34	T13	$(\bar{1}102)$	$[1\bar{1}01]$	
35	U14	$(\bar{1}012)$	$[10\bar{1}1]$	
36	V15	$(0\bar{1}12)$	$[01\bar{1}1]$	

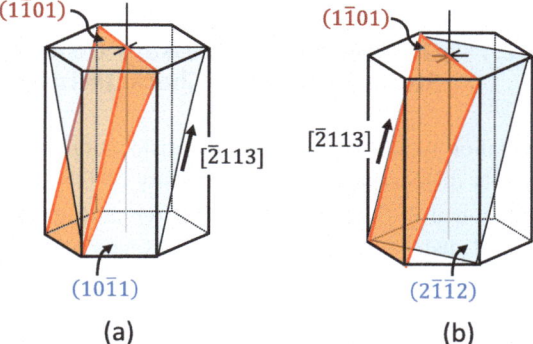

Fig. 4.10 Cross slips between pyramidal systems in HCP crystals. **a** Pyramidal 1 <a + c> $(10\bar{1}1)[\bar{2}113]$ and pyramidal 1 <a + c> $(1\bar{1}01)[\bar{2}113]$, **b** pyramidal 2 <a + c> $(2\bar{1}\bar{1}2)[\bar{2}113]$ and pyramidal 1 <a + c> $(1\bar{1}01)[\bar{2}113]$

4.1.4 An Extended Model for the Dislocation Mean Free Path

We have already modeled in Eq. (3.23) the contribution of dislocations accumulated in various slip systems to the mean free path of dislocations moving on a slip system. In order to represent the possibility that SS and GN dislocations play different roles in the mean free path, Eq. (3.23) could be modified as follows.

$$L^{(n)} = \frac{c^*}{\sqrt{\sum_m \left(w_S^{(nm)} \rho_S^{(m)} + w_G^{(nm)} \left\| \rho_G^{(m)} \right\| \right)}}, \quad (4.13)$$

where, $w_S^{(nm)}$ and $w_G^{(nm)}$ are the weight matrices for SS and GN dislocations accumulated in slip system m to the mean free path of dislocations in slip system n. These matrices are expected to have similarity to the slip interaction matrix $\Omega^{(nm)}$. For example, in FCC crystals, the weight matrices, $w_S^{(nm)}$ and $w_G^{(nm)}$ would be similar to that shown in Fig. 4.4. The weight component for the self- and coplanar- systems are related to elastic interaction, while the other components colored by white or gray should be related to their mutual cutting. Recalling that the stage I deformation was reproduced in good agreement with the experimental data when the mean free path was assumed constant, the component of the weight matrix corresponding to the elastic interaction should be small, especially for the matrix $w_S^{(nm)}$. On the other hand, the stress field formed by GN dislocations has a long ranging character because the sum of Burgers vectors is not equal to zero, and the weight matrix of $w_G^{(nm)}$ for GN dislocations may differ from that for SS dislocations, especially for the components representing the elastic interactions. However, detailed studies remain to be made.

4.2 Effect of Length Scale of Metal Microstructure on Strength and Strain Hardening

4.2.1 Geometrically Necessary Dislocations and Scale Dependent Strain Hardening Characteristics

Equation (4.13) assumes that the dislocation mean free path also depends on the density of GN dislocations. Since the density of GN dislocations is scale dependent, as already mentioned in Sect. 3.2, the mean free path given by Eq. (4.13) is scale dependent. The rate of dislocation accumulation becomes scale dependent by Eq. (3.10) and as a result, this is reflected in the calculated macroscopic stress–strain curves.

Analyses [13] were made using Eq. (4.13). The model used in the analysis is a Cu polycrystalline plate consisting of six crystal grains, as shown in Fig. 4.11a, with the mean grain sizes of 5, 1, or 0.2 μm. Crystal orientations of each grains are shown in Fig. 4.11b. Displacement of the bottom surface in y direction was fixed and a uniform tensile displacement was applied on the top surface.

Figure 4.12a–c show the distributions of the sum of dislocation density norm $\sum_n \|\rho_G^{(n)}\|$ when the nominal tensile strain is 5%. Results for the specimens with the mean grain size of 1 and 0.2 μm are shown in enlarged figure. Density of GN dislocations is higher near some grain boundaries, and other patterns with higher density are also observed inside the grains. Comparing the results for specimens with different mean grain sizes, density of GN dislocations is higher in smaller specimens.

Specimens with mean grain size ranging from 0.1 to 500 μm were made, where the shape and the crystal orientation of each grains are the same to that shown in Fig. 4.11a and b. Figure 4.13a shows the stress–strain curves obtained by tensile deformation analyses. We also performed the analysis with the condition $w_G^{(nm)} = 0$, where GN dislocations does not contribute to the dislocation mean free path, and denoted the result as scale independent model. The results show that the yield stress is not scale dependent, but the strain hardening properties are different for each specimen. This scale dependent nature of strain hardening is attributed to the fact that the density of GN dislocations is higher in smaller specimens, resulting in shorter mean free path and quicker accumulation of SS

4.2 Effect of Length Scale of Metal Microstructure on Strength and Strain ...

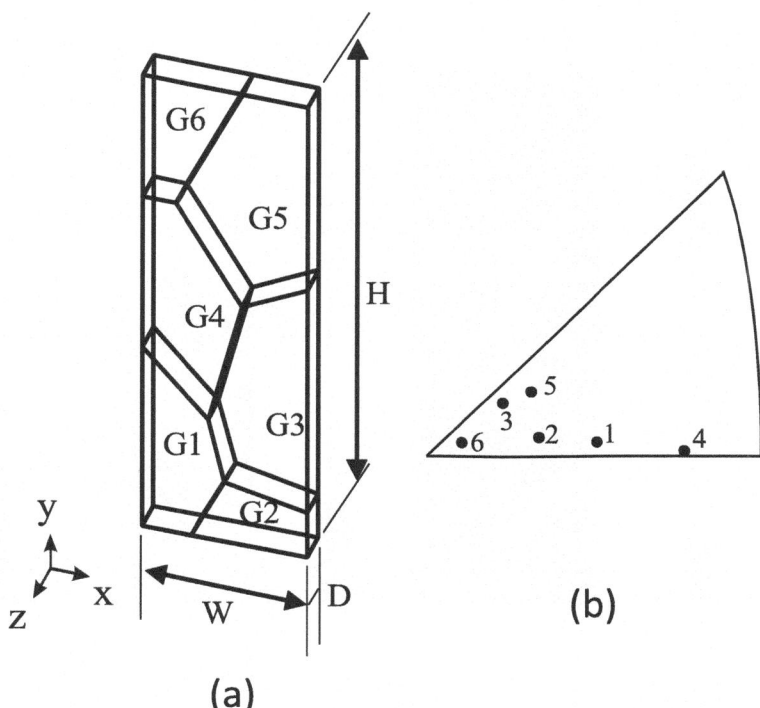

Fig. 4.11 a Geometry of the simplified polycrystal model employed for the analysis. The model consists of six grains and each grain penetrate the specimen in z-direction. The mean diameter of grains is approximately equal to the width of the specimen. Displacement in *y*-direction is constrained at the bottom surface, while uniform tensile displacement is given at the top surface. **b** Crystal orientation of the tensile axis of the six grains [13].

dislocations. Figure 4.13b shows the nominal tensile stress plotted as against the square root of the mean grain size. Compared to the linear character predicted by the Hall–Petch relationship, the obtained result shows a deviation from the linear one. However, if we observe the relationship in a narrower window, for example from 1 to 50 μm, an approximation that the plastic flow stress is proportional to the mean grain size holds.

4.2.2 Scale Dependent Modeling of the Yield Stress of Polycrystals

In Fig. 4.13a, the strain hardening was larger in specimens with smaller grain size, while the yield stress did not change. This is contradictory to experimentally observed facts. Let us discuss this subject.

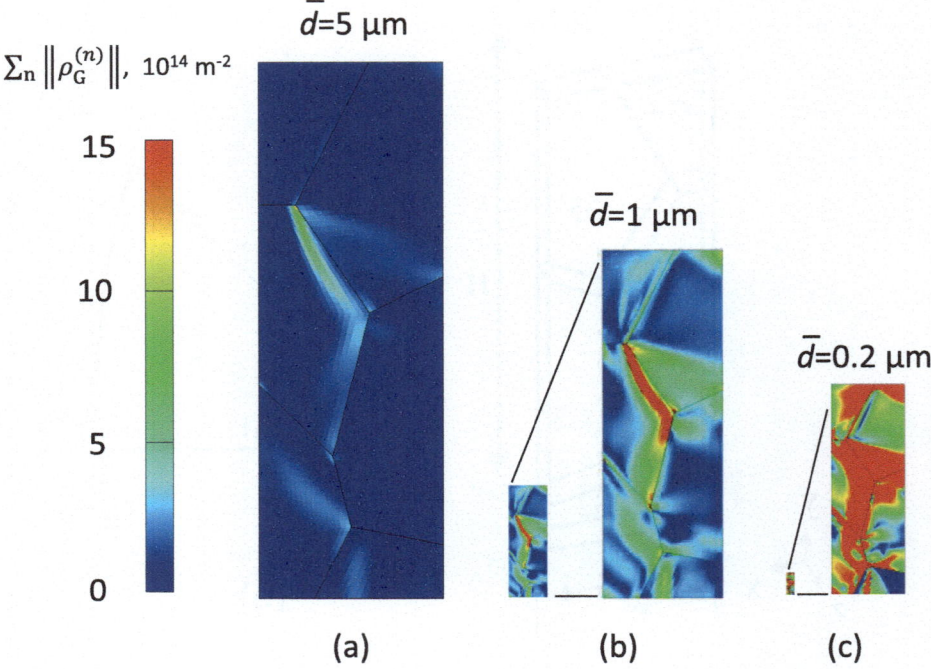

Fig. 4.12 Distribution of the density norm of GN dislocations in specimens with the mean grain size of 5, 1, and 0.2 μm. Results for smaller specimens ($\bar{d}$=1 and 0.2 μm) are shown enlarged for clarity. Nominal tensile strain is 5%. Data $c^* = 100$, $D = 0.5$ nm, $\rho_{0\_tot.} = 1.2 \times 10^{12}$ m^{-2} were used. The weight matrices $\mathbf{w_S}$ and $\mathbf{w_G}$ for SS and GN dislocations were assumed to be the same, and the components of the matrices for self- and coplanar interaction were set to zero, and the other components to 1. The interaction matrix Ω adopts a quasi-isotropic condition and all components are approximately equal to 1 [13]

Yielding is a phenomenon that the macroscopic stress–strain curve deviates significantly from the linear elastic line. The stress at which polycrystals exhibit such yielding, called as the yield stress, varies with the mean grain size and the relationship is called, in some cases, the Hall–Petch relationship. Yielding occurs by plastic deformation inside the material. However, many experimental studies have revealed that in FCC crystals, for example, fine slip lines form in the microstructure well before the macroscopic yielding occurs. Figure 4.14 shows the slip line observation before and after the macroscopic yielding [14]. In stage (a) of the load-elongation curve, slip lines are observed in some grains, and in stage (b), where macroscopic yielding occurs, slip lines are observed not only in large-sized grains but also in small-sized grains. This result depicts that plastic slip occurs before the macroscopic yield point in grains of larger size, and as deformation

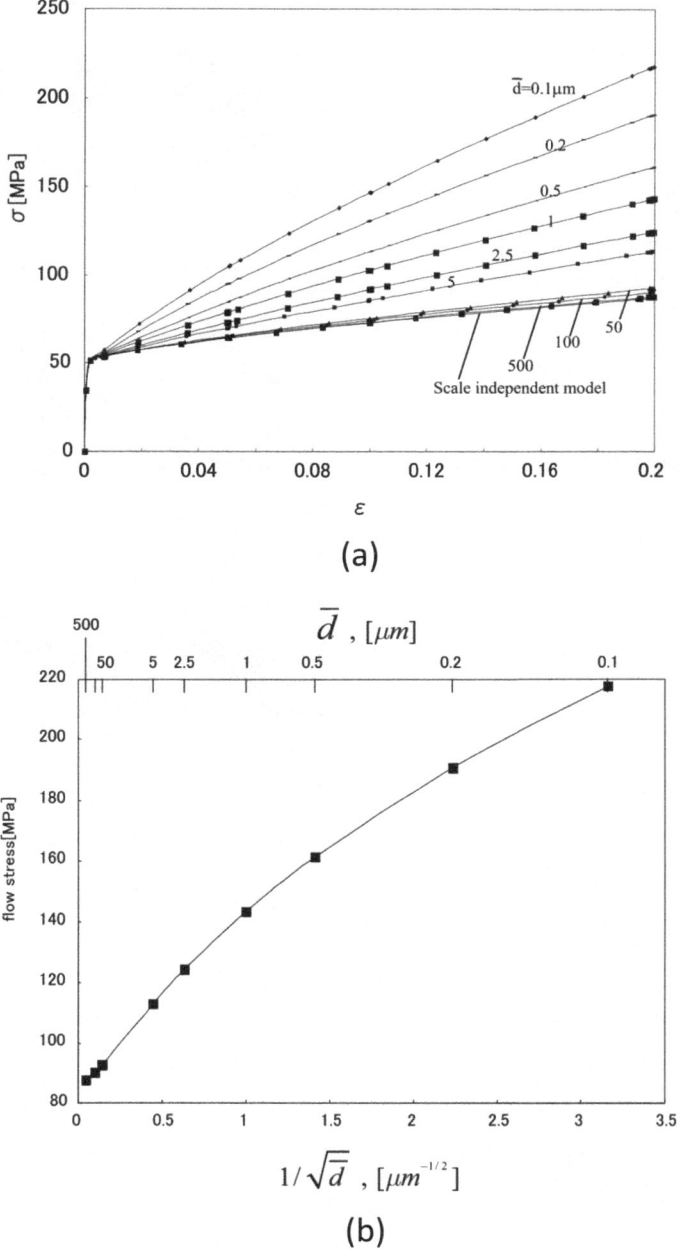

Fig. 4.13 a Nominal stress–strain curves of specimens with various mean grain size and, **b** Hall–Petch plot of the flow stresses when the mean tensile strain is about 20% [13]

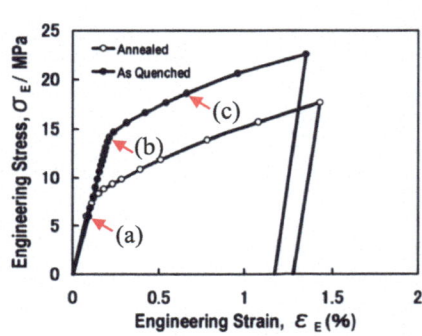
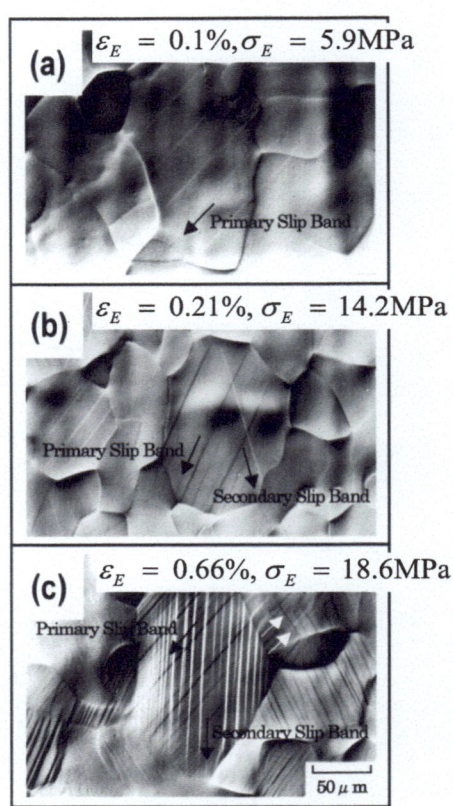

Fig. 4.14 Experimental observation of slip lines before and after the macroscopic yield point in a pure Al polycrystal plate [14]

proceed with increased external load, slip lines are observed also in grains of smaller size and slips on more than one system are superimposed.

Let us examine the reason why the slip deformation occurs first in grains of larger size and later in grains with smaller size. Except for ultrafine-grained materials, slip deformation in grains occurs by the action of Frank-Read dislocation sources. Using dislocation dynamics simulations, we examined [15, 16] the process of dislocation loop emission from a dislocation source in a crystal grain shown in Fig. 4.15. The model is a cube shaped crystal with lateral dimension d and the slip plane is placed across the center of the model. A Frank-Read source of length λ is introduced. The source is a dislocation segment of pure edge type and its ends are pinned. When d is sufficiently larger than λ, the stress needed to emit a dislocation loop is given by the Orowan stress of Eq. (2.5).

In order to mimic grain boundaries, we imposed a condition that dislocations cannot escape from the surfaces of the model. Simulation results when $d = 1.5$ μm and $\lambda =$

4.2 Effect of Length Scale of Metal Microstructure on Strength and Strain ...

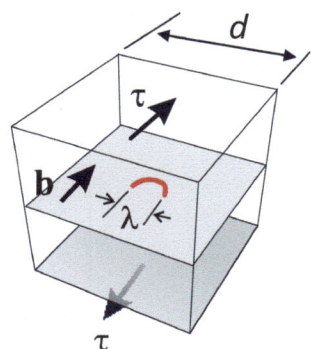

Fig. 4.15 Conditions used to simulate dislocation emission from a Frank-Read source in a confined media

0.5 μm (i.e. $d/\lambda = 3$) are shown in Fig. 4.16. A shear stress of approximately twice the Orowan stress is applied in the direction parallel to the Burgers vector.

Because of the apparent mass of dislocations, a finite amount of time is required for them to move under the application of stress. At the initial stage of movement, dislocation line bows out from the Frank-Read source (a), and then a part of dislocation line collides with the grain boundary (in this case, the outer surface of the specimen), where the collided part stops its movement (b) but the remaining parts continue moving. After further time, a closed dislocation loop and a new dislocation segment is formed where the Frank-Read source was, as shown in (d).

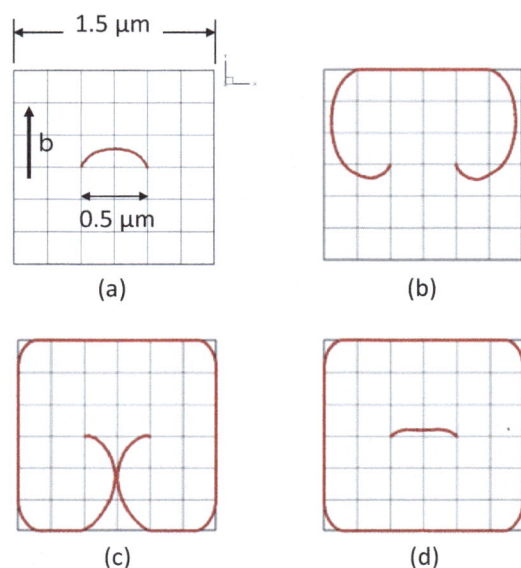

Fig. 4.16 Emission of a dislocation loop when $d = 1.5$ μm and $\lambda = 0.5$ μm. Under the application of shear stress, dislocation line starts to bow out (**a**) and after a time, some parts of dislocation line hit the specimen surface and stop their movement (**b** and **c**), but the remaining part continue to move and finally forms a closed loop (**d**)

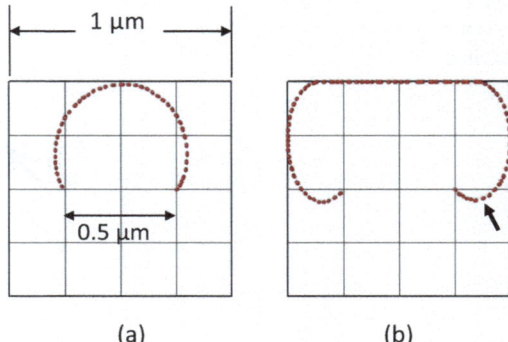

Fig. 4.17 Results of dislocation movement when $d = 1$ μm and $\lambda = 0.5$ μm. The dislocation arc initially expands freely (**a**) but stops moving at the state (**b**). The curvature of the part indicated by the arrow is so large that further movement is not possible due to the line tension

Figure 4.17 shows the results when $d = 1.0$ μm and $\lambda = 0.5$ μm (i.e. $d/\lambda = 2$). The simulation conditions other than the dimension of d are the same as those used in Fig. 4.16. In this case, dislocation line emitted from the Frank-Read source did not develop beyond the state shown in Fig. 4.16b even after a long period of time. This is because the curvature of the dislocation line increases at the part indicated by an arrow, and the forces due to line tension of dislocations and the applied stress are in equilibrium. If the applied shear stress is increased, for example, three times the Orowan stress τ_∞, this equilibrium is broken and an emission of closed dislocation loop is achieved.

These two results show that different grain sizes (in this case, d) can lead to the emission of a dislocation loop or stop midway, even with the same Frank-Read source length and applied stress. A quantitative study of this phenomenon was conducted and the results are shown in Fig. 4.18. The horizontal axis is the ratio d/λ, and the vertical axis is the ratio of stress needed to emit at least one dislocation loop within a finite time divided by the Orowan stress in the infinite body τ_{thres}/τ_∞. Here, the upper limit of the dislocation dynamics simulation steps was set to 750,000. It can be seen that the ratio τ_{thres}/τ_∞ is approximately a constant in the region of $d/\lambda > 3$, while the ratio shows a steep increase in the region of $d/\lambda \leq 3$. This means that when the grain diameter is larger than three times the dislocation source length, the stress required to emit a closed dislocation loop is approximately a constant, while the stress increases rapidly with the decrease of grain size. Figure 4.19 shows simulation result for the case when the dislocation source is placed off the center of the specimen. When the dislocation source is not at the center of the specimen, curvature of a part of dislocation line increases at an earlier stage and a larger shear stress is required to accomplish the formation of closed loop.

The Orowan stress given by Eq. (2.5) for the emission of dislocation loops in infinite body shows that larger the dislocation source length, smaller is the stress required to emit

Fig. 4.18 Ratio of the stress needed to form a closed dislocation loop and the Orowan stress, τ_{thres}/τ_∞ plotted against d/λ. When $d/\lambda < 3$, τ_{thres}/τ_∞ increase rapidly [15, 16]

Fig. 4.19 a and b Result of the dislocation dynamics simulation when the Frank-Read source is placed near the grain boundary. $d = 1\,\mu m$, $\lambda = 0.2\,\mu m$ are used and the applied stress is $\tau = 1.02\tau_\infty$. The movement of dislocation arc stops before the formation of closed loop [16].

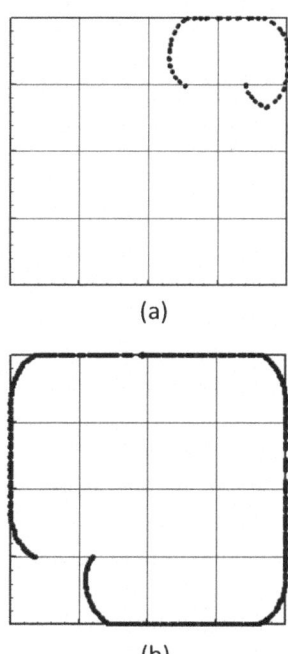

the loop. When the loop emission occurs in a crystal grain of finite size, another length scale contributes to the dislocation loop emission; when the grain size is small, the stress required to emit a dislocation loop increases. The results obtained in Fig. 4.18 shows that the optimum size of dislocation source in a crystal grain of size d is given by $\lambda \approx d/3$ and placed at the center of the grain. Therefore, if we assume that there are dislocation sources of any size smaller than the grain size, the minimum stress required to generate a closed dislocation loop is given by,

$$\tau_{loff} \approx \beta \frac{\mu b}{\left(d/3\right)} = 3\beta \frac{\mu b}{d}. \tag{4.14}$$

In the region of $d/\lambda \geq 3$ in Fig. 4.18, τ_{thres}/τ_∞ is larger for larger λ, showing that there is another scale dependency. The parameter β represents this effect in Eq. (4.14). Scale dependency of β was examined, but it was not significant within the length scale studied in the above-mentioned simulations and β is considered to be a constant here. However, in ultrafine-grained materials with grain diameters of 1 μm or smaller, the Frank-Read sources is unlikely to exist within grains, and dislocations are supplied from the grain boundaries. It has been shown that the stress required for dislocation emission in that case is proportional to $3\mu b/d \times \log(d/b)$, which is approximately a few times larger than that given by Eq. (4.14). In this article, we do not go into the details of grain refinement and the stress required for dislocation emission is given by Eq. (4.14).

There are various ways to incorporate the stress τ_{loff} required to initiate plastic slip in a finite medium into the CRSS model, and a simple approach is as follows where the stress given by Eq. (4.14) is added to the lattice friction and Taylor hardening terms,

$$\theta^{(n)} = \theta_0^{(n)}(T) + \sum_{m=1}^{N} \Omega^{(nm)} a \mu b^{(m)} \sqrt{\rho_S^{(m)}} + 3\beta \frac{\mu b}{d^{(n)}}. \tag{4.15}$$

Let us call the third term in the right-hand side of Eq. (4.15) as the Orowan stress term or Orowan stress term in a finite medium. Figure 4.20 shows macroscopic stress–strain relationship obtained for the models shown in Fig. 4.11a with mean grain size of 1, 5, and 50 μm and the model for CRSS given by Eq. (4.15). The smaller the mean grain size, the larger the macroscopic yield stress.

Grain boundary is not the only source of the scale dependence but phase boundaries, secondary particles dispersed in the microstructure and other factors also contribute to the effect. Extending the meaning of d in Eq. (4.15), it can be considered as a quantity representative of the microstructural length scale that control the movement of dislocations. For example, in the case of alloy steels dispersed with hard precipitates, distance between precipitates will be one of such length scales. In the case of slip deformation in ferrite lamellae in pearlite steels, movement of dislocations is blocked at the ferrite-cementite phase boundaries. The distance between phase boundaries when the ferrite layer is cut by a slip plane can be the representative length of the microstructure and this distance varies from slip system to slip system depending on the crystal orientation and the direction of boundaries. Quantitative approaches of these subjects are introduced in Sect. 4.2.4.

Fig. 4.20 Load-elongation curves obtained for tensile deformation of the polycrystal model shown in Fig. 4.10 and applying the model for the critical resolved shear stress given by Eq. (4.15), assuming $\beta = 1$, $d^{(n)} = \bar{d} \, (= w)$. Macroscopic yield stress depends on the mean grain size [15, 16]

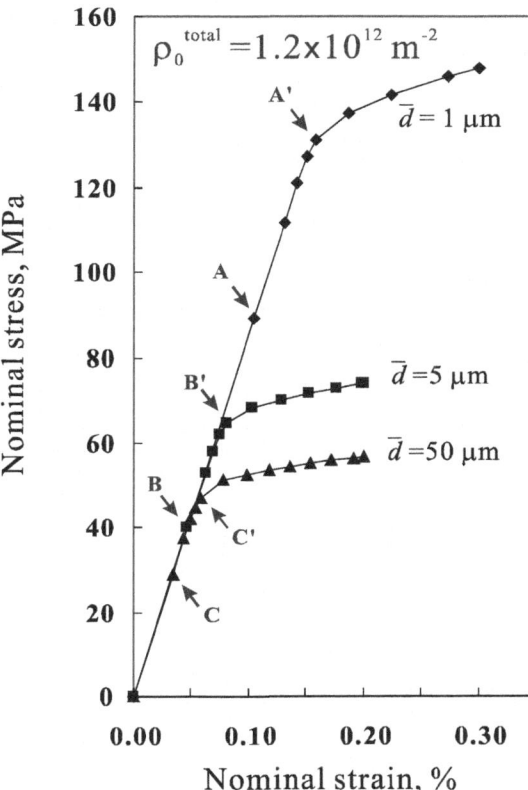

4.2.3 Deformation Twin and Critical Resolved Shear Stress

Beyerlein and Tomé [17] studied plastic slip and twinning deformations of Zr with HCP lattice structure in terms of dislocations, grain boundaries and twin interfaces. The main points of their study on the interaction between slip and twinning are summarized as follows.

When slip deformation occur in the space where twinning already took place, twin boundaries act as a strong obstacle to slip. If the CRSS for the slip system n is written in the similar form as Eq. (4.15), the obstacle effect is, according to the authors, given by the third term of the following equation,

$$\theta^{(n)} = \theta_0^{(n)}(T) + \sum_{m=1}^{N} \Omega^{(nm)} a\mu b^{(m)} \sqrt{\rho_S^{(m)}} + \kappa\mu \left(\frac{b^{(n)}}{d_{twin}^{(n)}}\right)^q, \qquad (4.16)$$

where, $q = 1/2$ in their study. The third term gives the scale dependent character of the Hall–Petch type and the coefficient κ is considered to be determined by the geometry of dislocations moving in the region between twinned regions. The length scale $d_{twin}^{(n)}$ develops with the introduction and growth of twinned region and can be quantified as follows.

It is assumed that crystal grains are divided into twin regions and matrix when the calculated volume fraction of twinned region reaches a predetermined value. The twinned region is flat and the thickness of the twinned region increases as deformation proceeds. Nucleation of twin is not treated explicitly, and when the ratio of the calculated twin strain to the crystallographically determined twin strain reaches 5%, the twin system is selected as the predominant twin system in the grain. As shown in Fig. 4.21, twinned regions are arranged so that layers with the same thickness d^t are stacked at equal intervals d^c in the grain. The controlling length scale $d_{twin}^{(n)}$ is the distance of two intersections formed when a slip plane of slip system n intersects the twin interfaces, and given by,

$$d_{twin}^{(n)} \sin \varphi = d^c - d^t, \qquad (4.17)$$

where, φ is the angle formed by two normal vectors of the slip plane and the twin boundary plane.

Until the twin strain calculated by the crystal plasticity analysis reaches a certain value, i.e., $\gamma_{twin,0}$, the twin is considered to be in the nucleation stage and no interface is introduced into the analysis. For $\gamma_{twin} > \gamma_{twin,0}$, the volume fraction V_f of the twinned region is,

$$V_f = \frac{d^t}{d^c}, \qquad (4.18)$$

where, Beyerlein and Tomé [17] assumes that d^c is a constant (e.g. 0.2 times the grain size for the tensile twin in Zr). Assuming that the crystallographically determined twin

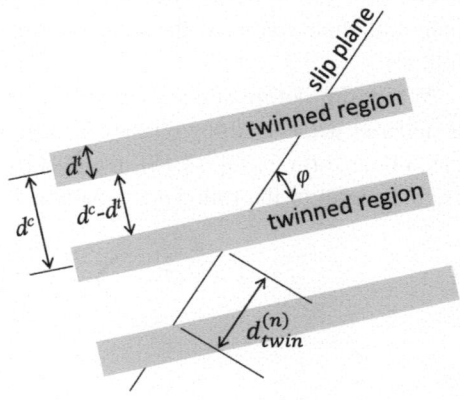

Fig. 4.21 Microscopic configurations and length scales to consider when deformation on slip system n takes place in a medium with repeated layers of twinned regions

4.2 Effect of Length Scale of Metal Microstructure on Strength and Strain ...

strain is Γ_{twin} and the maximum volume fraction of twinned region in the crystal grain is $V_{f,max}$, the volume fraction of twinned region V_f is given as follows,

$$V_f = \frac{\gamma_{twin}}{\Gamma_{twin}} V_{f,max}. \tag{4.19}$$

From Eqs. (4.17)–(4.19), the length scale is determined as follows.

$$d_{twin}^{(n)} = d^c \left(1 - \frac{\gamma_{twin}}{\Gamma_{twin}} V_{f,max}\right) / \sin\varphi. \tag{4.20}$$

From TEM observation, Beyerlein and Tomé [17] assumes $V_{f,max} = 0.5$. Therefore, of the quantities on the right-hand side of Eq. (4.20), γ_{twin} is the only variable that evolve during deformation. Evaluating this value in the crystal plasticity analysis allows us to incorporate microscopic length scale due to the presence of twinning interfaces.

The resistance to the propagation of the twin deformation in the region where slip has already occurred can be written as,

$$\theta^{(n)} = \theta_0^{(n)} + \mu \sum_m C^{(nm)} b^{(n)} b^{(m)} \rho^{(m)} + \sum_{m \neq n} \theta_{\text{interface}}^{(m)} + \kappa' \mu \left(\frac{b^{(n)}}{d^{(n)}}\right)^q. \tag{4.21}$$

Here, the first term $\theta_0^{(n)}$ is the stress required for twin nucleation and the friction stress for its propagation after the nucleus is developed. Nucleation is considered to occur as a stochastic process through the thermal activation process. The second term is the effect of dislocations introduced into the microstructure. Possibilities that preexisting dislocations play roles in the formation and propagation of twinning has been discussed [18]. When $C^{(nm)} > 0$, dislocations act as obstacles to the propagation of twin, while $C^{(nm)} < 0$ is also possible, where dislocations promote the nucleation and propagation of twin.

The third term gives the propagation resistance from the twin boundary introduced by twin system m, which is different from the twin system n. The form of the mathematical function is the same as that for the fourth term, and the length scale is the distance that the twin system n passes through the interface of the twin system m as it propagates. Authors [17] assumed that the dimensionality dependence of the third and the fourth terms are proportional to $1/\sqrt{(\text{length scale})}$, but this would depend on the physical details of the twin deformation as they penetrates the interface. There could also be a more detailed discussion of the value of the length scale.

4.2.4 Introduction of Microstructural Length Scale into the Model of the Dislocation Mean Free Path

In the previous section, various length scales established in the microstructure were shown to play important roles in the resistance to the slip or twin deformation which originate from the interactions between the deformation modes. As will be anticipated, these length scales not only control deformation resistance, but also have an effect on strain-hardening properties through the mean free path of dislocations. Let us model the dislocation mean free path incorporating the length scale of microstructure as follows,

$$L^{(n)} = Min\left[\frac{c^*}{\sqrt{\sum_m \left(w_S^{(nm)} \rho_S^{(m)} + w_G^{(nm)} \left\|\rho_G^{(m)}\right\|\right)}}, d^{*(n)}\right], \quad (4.22)$$

where, the function Min selects the smallest value among the arguments and $d^{*(n)}$ denotes the length scale observed in deformation mode (i.e. slip or twin system) n. The first argument is the effective mean spacing of dislocations accumulated in the microstructure and when this distance is smaller than the microstructure length scale $d^{*(n)}$, then the mean free path is determined by the density and type of accumulated dislocations. On the other hand, if the microstructure length scale is smaller than the effective mean spacing of accumulated dislocations, then the mean free path is given by the microstructure length scale.

Although there are still many points to consider in determining the controlling microstructure length scales, some attempts have already been made. In the study of macroscopic stress–strain curves of alloy steels with heterogeneous precipitation of fine hard particles, Okuyama et al. [19] made a numerical model of such an alloy, determined the inter-particle spacings at different locations in the material and adopted these values as the length scales in the models of CRSS given by Eq. (4.15) and mean free path given by Eq. (4.22).

Figure 4.22 shows the model of the microstructure where the average diameter of the particles is 39 nm and the volume fraction is 1.24%. Focusing on a certain point in the material, a region of thin plate parallel to the slip plane and including the point is imaginarily cut out as shown in Fig. 4.22b. Among the particles included in the plate, particles in the neighborhood of the point of interest are extracted and distances between them are calculated as shown in Fig. 4.22c. Figure 4.22d shows the distribution of the mean spacing of particles in a plane parallel to the x–y plane. Figure 4.22e shows the stress–strain curves obtained from the crystal plasticity analyses. In addition to the mean spacing of particles as a microstructure length scale, the minimum and maximum values of particle spacing were also used for comparison. As compared in Fig. 4.22e, analysis results employing the mean or maximum value of particle spacing as the microstructure length scale are in good agreement with the experimental results.

Yasuda et al. [20] determined the length scales used in Eqs. (4.15) and (4.22) for the slip deformation in pearlite microstructure by considering the geometric relationship of ferrite-cementite interface, slip plane and slip direction. Assuming that slip deformation occurs in the ferrite layer sandwiched between cementite layers as shown in Fig. 4.23a, the controlling length scale for the expansion of the Frank-Read type dislocation loops is $\hat{d}/2$ or $\hat{d}/3$, where $\hat{d}$ is the distance of intersections formed by the slip plane and

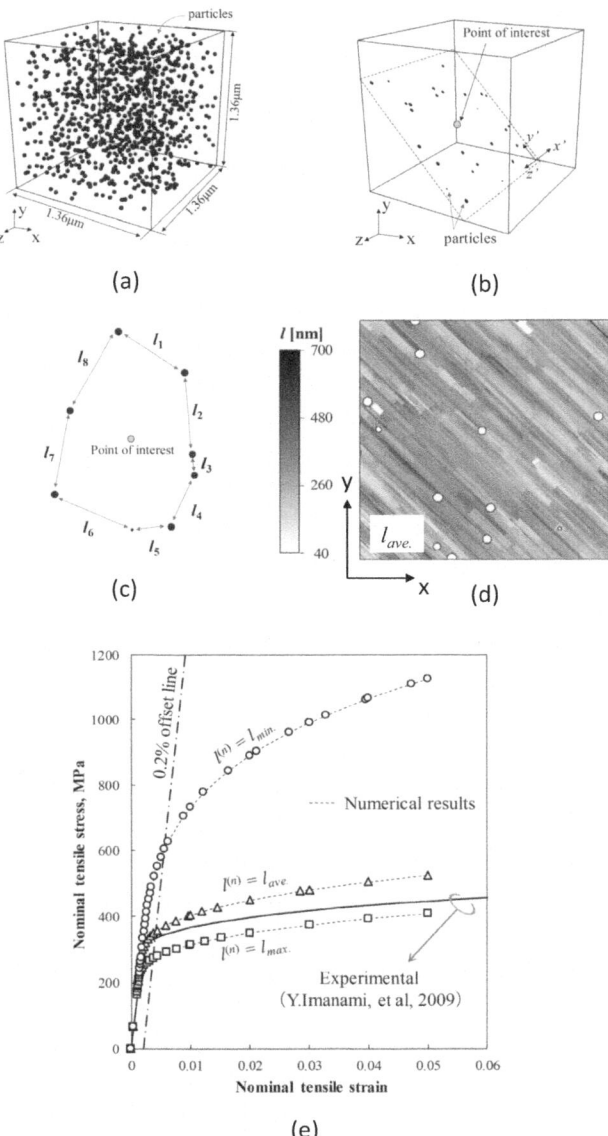

Fig. 4.22 Analysis of precipitation-hardened steel. **a** Model of the microstructure, **b** non-uniformly distributed precipitates on a slip plane, **c** distance between precipitates around the point of interest, **d** distribution of average distance between precipitates, and **e** macroscopic stress–strain relationship obtained using minimum, average and maximum distance between precipitates as the microscopic length scale [19]

ferrite-cementite interfaces. On the other hand, the distance of two points where a line parallel to the Burgers vector penetrates the ferrite-cementite interfaces is $d*$ and this is considered as the dislocation mean free path. Both length scales can be determined from the geometry of the ferrite-cementite interface and the arrangement of the slip system.

Figure 4.23b shows the results of stress–strain relationship analysis of the ferrite layer thicknesses of $d = 50$ and 500 nm. The Bagaryatsky or Pitsch-Petch orientation relationship was adopted for the ferrite and cementite layers. The stress–strain curve labeled ferrite5 in the figure is an artificial one and serves as a guide for the strain-hardening properties necessary to suppress plastic instability in the pearlite microstructure [21]. When $d = 50$ nm, the strain hardening of the ferrite layer is larger than that of the ferrite5 in

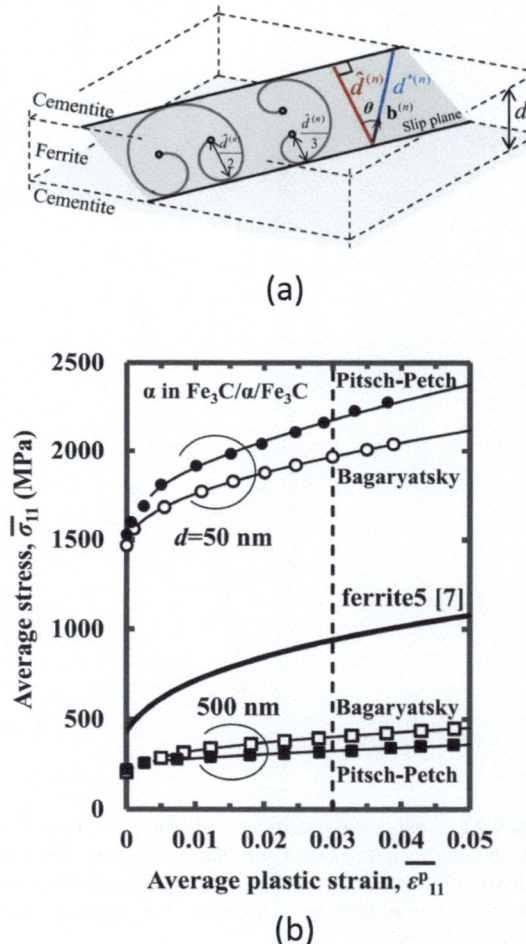

Fig. 4.23 Stress–strain relationship for ferrite layers in pearlite microstructure. The microscopic length scales $\hat{d}$ and $d*$ were used in Eqs. (4.15) and (4.20) and Bagaryatsky or Pitsch-Petch orientation relationship was introduced for the ferrite and cementite regions. The results show that for a ferrite thickness of 50 nm, strain hardening rate is high enough to suppress unstable deformation of the cementite layer [20]

both cases with Bagaryatsky or Pitsch-Petch relationship and the unstable and localized deformation of cementite layers is suppressed, at least initially.

Shimokawa et al. [22] made a molecular dynamics simulation study of interaction between a <112> asymmetric tilt grain boundary and an edge dislocation. Before the introduction of slip dislocations, there were intrinsic grain boundary dislocations in the boundary. When an edge dislocation was introduced and pushed to the grain boundary, the intrinsic boundary dislocations rearranged themselves and the externally supplied dislocation was absorbed to the grain boundary to form an extrinsic boundary dislocation.

Absorption of dislocations into grain or phase boundaries is not explicitly modeled in Eq. (4.22), but it would be possible to include the effect approximately by setting $d^{*(n)}$ larger than the distance determined geometrically. Yasuda et al. [20] have made a preliminary study on this point by setting $d^{*(n)}$ five times larger than the geometrically determined distance and found that the strain-hardening rate of ferrite layer decreases, that is, even if the thickness of the ferrite layer does not change, the deformation of cementite is instable when the lattice dislocations are absorbed into the phase boundary. There are a lot of interesting subjects to be studied further.

The models given by Eqs. (4.16) and (4.21) show that the twin boundaries play a significant role in the resistance to plastic deformation. Expanding this idea, it is natural to assume that twin boundaries also contribute to the dislocation mean free path. Furthermore, and not limited to the twin boundaries, the length scales that are considered to contribute to the movement of dislocations can be expressed as $d_1^{*(n)}, d_2^{*(n)}$.... Then, a more generalized expression for the mean free path could be,

$$L^{(n)} = Min\left[\frac{c^*}{\sqrt{\sum_m \left(w_S^{(nm)}\rho_S^{(m)} + w_G^{(nm)}\left\|\rho_G^{(m)}\right\|\right)}}, d_1^{*(n)}, d_2^{*(n)}...\right]. \qquad (4.23)$$

Length scales, $d_1^{*(n)}, d_2^{*(n)}$... can be constants during deformation as in grain size or precipitate spacing, or they can vary with deformation as in the distance between twin boundaries.

References

1. Cottrell AH (1952) Dislocations and plastic flow in crystals, 1st editio. Oxford University Press, Oxford
2. Narita N (1985) Bull Japan Inst Met 21:984
3. Kocks UF, Mecking H (2003) Prog Mater Sci 48:171
4. Mecking H, Kocks UF (1981) Acta Metall 29:1865
5. Kocks UF. Constitutive Behavior Based on Crystal Plasiticity, in:. Miller AK (Ed.). Unified Const. Equations Creep Plast. Springer Netherlands; 1987.
6. Estrin Y, Mecking H (1984) Acta Metall 32:57

7. Hirth JP, Lothe J. Theory of Dislocations, second ed. New York. US.: John Wiley & Sons; 1982.
8. Ohashi T (1994) Phil Mag A 70:793
9. Ohashi T (1987) Trans JIM 28:906
10. Jackson PJ, Basinski ZS (1967) Can J Phys 45:707
11. Franciosi P, Berveiller M, Zaoui A (1980) Acta Metall 28:273
12. Ando S, Takashima K, Tonda H (1990) J Japan Inst Met 54:427
13. Ohashi T. A New Model of Scale Dependent Crystal Plasticity Analysis, in:. Kitagawa H, Shibutani Y (Eds.). IUTAM Symp. Mesoscopic Dyn. Fract. Process Mater. Strength - Solid Mech. Its Appl. Vol. 115 -. Osaka, Japan: Kluwer Academic Press; 2003.
14. Ono N, Hayakawa S, Miura S (2002) Trans JSME Ser A (in Japanese) 68:1129
15. Ohashi T, Kawamukai M, Zbib H. Crystal Plasticity Modeling of Scale Dependency of Yield Stress for Metal Polycrystals, in:. Khan AS, Kazmi R (Eds.). Proc. Plast. '06 Twelfth Int. Symp. Plast. Its Curr. Appl. Ed. Halifax: NEAT Press; 2006.
16. Ohashi T, Kawamukai M, Zbib H (2007) Int J Plast 23:897
17. Beyerlein IJ, Tomé CN (2008) Int J Plast 24:867
18. Yoo MH, Loh BTM. Structural and Elastic Properties of Zonal Twin Dislocations in Anisotropic Crystals, in:. Simmons J, De Wit R, Bullough R (Eds.). Fundam. Asp. Dislocation Theory. Nat. Bur. Stand.; 1970.
19. Okuyama Y, Tanaka M, Ohashi T, Morikawa T (2020) ISIJ Int 60:1819
20. Yasuda Y, Ohashi T, Shimokawa T, Niiyama T (2017) Mater Sci Technol 0836:1
21. Ohashi T, Roslan L, Takahashi K, Shimokawa T, Tanaka M, Higashida K (2013) Mater Sci Eng A 588:214
22. Shimokawa T, Kinari T, Shintaku S (2007) Phys Rev B - Condens Matter Mater Phys 75:1

Generation of Atomic Vacancies by Dislocation Pair Annihilation

5

Abstract

Density of atomic vacancies increases rapidly during plastic deformation and these vacancies play roles in diverse aspects of the physical properties of metallic crystals. This chapter reviews a theory of the density evolution of atomic vacancies due to pair annihilation of dislocations during plastic slip deformation and proposes an extended model that is compatible with three-dimensional crystal plasticity analysis. Results of some numerical analysis are presented, and it is shown that the vacancy density is significantly high when slips on more than one slip systems occur. A possible modification to the model of critical resolved shear stress with increased density of atomic vacancies is presented.

Keywords

Dislocation pair annihilation • Vacancy formation and sweep • Rate of vacancy formation • Rate of vacancy sweep

Atomic vacancy is a lattice defect where an atom that should be present at a lattice site is absent. Thermal equilibrium concentration [1] of vacancies derived by thermodynamics is given by [2],

$$c = exp\left(-\frac{h_v^f}{kT}\right), \tag{5.1}$$

where, k, T and h_v^f denote the Boltzmann constant, absolute temperature and the formation enthalpy of atomic vacancies, respectively. Using Eq. (5.1), the vacancy concentration, for example, in iron (Fe) at 300 K is calculated to be approximately 6×10^{-26}, while the concentration after a tensile deformation up to 20% strain was reported [3] to

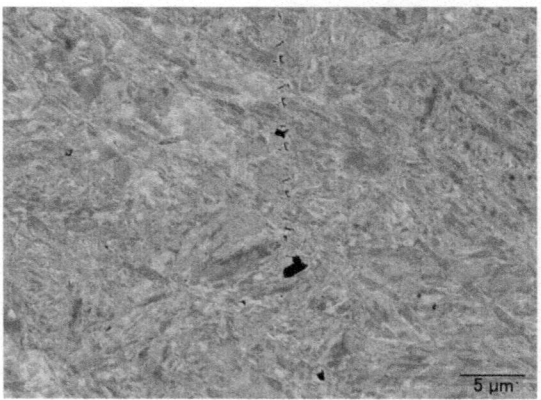

Fig. 5.1 Array of micro-meter sized voids observed near the fracture surface of martensitic steel

be on the order of 10^{-5}. Whatever the formation mechanism is, the vacancy concentration increases quite rapidly with plastic deformation.

There are a number of studies on the relationship between atomic vacancies and physical properties of metallic materials including, diffusion [4, 5], precipitation [6], recrystallization [7], creep [8], formation of nano-voids [9, 10], persisitent slip band and formation of fatigue cracks [11, 12], deformation localization and plastic instability [13], hydorgen embrittlement [14]. Figure 5.1 shows a SEM micrograph of a series of micron-sized voids observed near the fracture surface of martensitic steel (Tanaka: private communication, 2022). Large plastic deformation during the fracture process and the increase in vacancy density are assumed to contribute to the generation and development of these voids. To address such subject, we need to study in detail the generation and increase of atomic vacancies in microstructure.

Exprimental measurements of atomic vacancies include electrical conductivity, positron annihilation time, X-ray micro-diffraction and other techniques, but the spatial resolution of the data obtained is not yet high. Two mechanisms of the formation of vacancies due to plastic deformation are well known; one is the dragging of dislocation jogs and the other is the pair annihilation of edge dislocations. In an attempt to theoretically determine the evolution of atomic vacancy density with slip deformation, Cuitiño and Ortiz [9] discussed the drag mechanism of dilocation jogs and Essmann and Mughrabi [15] developed a theory of vacancy density evolution associated with pair annihilation of edge dislocations. In the following, the latter theory is introduced and implemented to the analysis of crystal plasticity and examine the development of vacancy density.

Figure 5.2 schematically shows the situation where two edge dislocation of opposite sign encounter [16]. There are cases where extra half-planes of edge dislocations meet in overlapping manner as shown in Fig. 5.2a and cases where there is a distance between the two extra-half planes as shown in Fig. 5.2b. In the former case, more than one atoms cannot coexist at the same lattice site and the extra atoms are emitted into the matrix

Fig. 5.2 Dipole formation of positive and negative edge dislocations when **a** the extra half-planes overlap or, **b** a row of vacancies is formed. (Redrawn after Kimura [16])

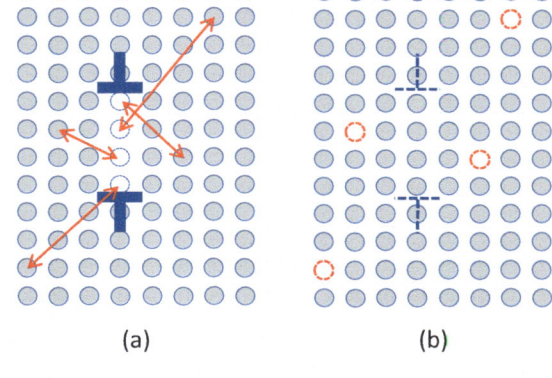

Fig. 5.3 **a** Vacancy type dipole of edge dislocations and, **b** its annihilation due to emission of atomic vacancies

and become interstitial atoms. In the latter case, vacant lattice sites between the extra-half planes are filled with atoms supplied from the matrix, but instead, atomic vacancies are left in the matrix as shown in Fig. 5.3. This process can be understood that pair annihilation of edge dislocations emitts atomic vacancies into matrix.

There is an approximate relationship between the formation enthalpies of interstitial atoms and vacancies as follows [1],

$$2h_v^f \approx \frac{1}{2}h_i^f. \tag{5.2}$$

Since the energy of interstitial atoms is a few times higher than that of vacancies, as shown in Eq. (5.2), interstitials are absorbed into sinks rapidly and vacancies remain longer.

As discussed in Sect. 3.2, Burgers vector of GN dislocations is deviated to positive or negative and the dislocations are not the mixture of positive and negative ones. Therefore, dislocations in an aggregate evaluated as geometrically necessary do not have their counterpart to annihilate each other. This means that the GN dislocations do not contribute to the generation of atomic vacancy or interstitials. Also, since screw dislocations are not associated with volmetric strain, they do not contribute to the generation of vacancies or interstitials even if pair annihilations occur. Let us consider the pair annihilation of edge

Fig. 5.4 **a** The number of vacancies emitted by the annihilation of a dipole of edge dislocations is proportional to D/b, **b** atomic vacancies in the range of y_p above and below the slip plane are swept by the motion of a dislocation

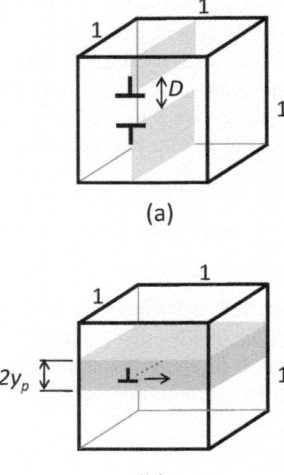

dislocations in SS dislocations. Pair annihilation has already been modeled by Eq. (3.7). For edge dislocations, Eq. (3.7) can be modified as follows,

$$d\rho^-_{S,\,edge} = \frac{D}{b}\rho_{S,edge}2c_e d\gamma, \qquad (5.3)$$

where $\rho_{S,\,edge}$ is the density of edge component among SS dislocations, and c_e denotes the ratio of contribution of edge dislocation component to the plastic shear strain $d\gamma$ and if the slip deformation is caused by an expansion of rectangular dislocation loops, as shown in Fig. 3.1, $c_e = 0.5$.

Since dislocations annihilate in pairs, the number of annihilation events during the plastic shear strain $d\gamma$ is,

$$d\rho^-_{S,\,edge}/2. \qquad (5.4)$$

Suppose there are a pair of positive and negative edge dislocations separated by a distance D in a cube shaped region of size $1 \times 1 \times 1$ as shown in Fig. 5.4a. The number of lattice sites counted in the direction of the distance between dislocations is D/b and the number counted in the direction of the dislocation line is $1/b$. Therefore, the number of vacancies generated by the annihilation of this pair is,

$$(D/b) \times (1/b). \qquad (5.5)$$

Since the number of annihilation events is given by Eq. (5.4), the number of vacancies emitted per unit volume is given by,

5 Generation of Atomic Vacancies by Dislocation Pair Annihilation

$$dN^+ = \frac{D}{2b^2} \frac{d\rho^-_{S,edge}}{2}. \tag{5.6}$$

The number of vacancies per unit volume (= vacancy density) is denoted by N and here, dN^+ is its increment. The number of annihilation events given by Eq. (5.4) includes both the cases of vacancy generation or interstitial generation. In Eq. (5.6), half of the annihilation event contribute to the generation of vacancies. Assuming that the atomic density is denoted by Z and $Z \approx 1/b^3$, Eq. (5.6) is written as follows.

$$dN^+ = \frac{bD}{2} Z \frac{d\rho^-_{S,edge}}{2}. \tag{5.7}$$

Substituting Eq. (5.3) into Eq. (5.7), we obtain,

$$dN^+ = \frac{ZD^2 c_e}{2} \rho_{S,edge} d\gamma. \tag{5.8}$$

Next, the amount of vacancies swept by moving dislocations is evaluated. Consider that when one dislocation passes through, it sweeps vacancies within $2\, y_p$ above and below the plane it passes through, as shown in Fig. 5.4b. The volume of the region where this sweep occurs is $1 \times 1 \times 2y_p$ and the number of vacancies in this volume is given by $2Ny_p$. Because the plastic shear strain caused by the passage of n dislocations is $d\gamma = n \times b/1$, the number of vacancies swept out when n dislocations pass through is given by,

$$dN^- = 2Ny_p(c_e d\gamma/b). \tag{5.9}$$

The strain increment on the right hand side of Eq. (5.9) is given by $c_e \times d\gamma$ instead of $d\gamma$, expressing also that the contribution of edge component of the dislocations is extracted. From Eqs. (5.8) and (5.9), the increment of the vacancy density is given by,

$$dN = \left(\frac{ZD^2 c_e}{2} \rho_{S,edge} - \frac{2y_p c_e}{b} N \right) d\gamma. \tag{5.10}$$

To utilize the result for the evolution of vacancy density of Eq. (5.10) in the crystal plasiticity analysis, the increments of vacancy densities caused by the activity of each slip systems are evaluated and summed. We used the folloing models [17, 18],

$$dN^{(n)} = \left(\frac{ZD^2 c_e}{2} \rho^{(n)}_{S,edge} - \frac{2y_p c_e}{b} N^{Tot} \right) d\gamma^{(n)}, \tag{5.11}$$

$$dN^{Tot} = \sum_n dN^{(n)}, \tag{5.12}$$

$$N^{Tot} = \int dN^{Tot}, \tag{5.13}$$

where, $dN^{(n)}$ is the increment of the vacancy density resulting from the activity of the slip system n and N^{Tot} is the total density. When determining the density of SS dislocations in Eq. (3.9), we did not distinguish between edge and screw dislocations but simply determined them as "density". To extract the edge component from the density of SS dislocations, we can use the aspect ratio α of dislocation loops. The following,

$$\rho_{S,edge} = \rho_S/(1+\alpha), \tag{5.14}$$

is used to evaluate the density of the edge component of the SS dislocations.

Equations (5.11)–(5.14) were used to analyze the change in density of atomic vacancies when a slip deformation occurs in a single crystal material. The representative length scale of the specimens is $d = 1$ mm, which is considered large enough, the initial dislocation density is $\rho_0 = 1.2 \times 10^{10}$ m^{-2}, which is considered low enough, and the lattice friction stress is $\theta_0 = 0.3$ MPa. The annihilation distance of the dislocations was set to $D = 5b$, the range of vacancy sweep by the dislocations was set to $y_p = b$, and $c^* = 100$ so that the interaction between the forest dislocations and the moving dislocations was also very weak. Input data used in the analysis are given in Table 5.1 and results are shown in Figs. 5.5 and 5.6.

Figure 5.5a shows the load-elongation curve of the specimen oriented for single slip. After parabolic strain hardening, the strain hardening rate approaches to zero when the nominal tensile strain is about 0.2. This is due to the saturation of dislocation density caused by the equilibrium between the accumulation and annihilation terms of SS dislocations in Eq. (3.9). Figure 5.5b plots the rates of vacancy formation and sweep given by Eqs. (5.8) and (5.9). As the SS dislocation density saturates, the rate of vacancy formation becomes constant, while the rate of vacancy sweep increases. Figure 5.5c shows the evolution of the vacancy density. The curve is S-shaped and at plastic shear strain 1, the vacancy density is approximately 1.5×10^{22} m^{-3}, or about $0.25 \times 10^{-6} = 0.25$ ppm in concentration.

Figure 5.6 compares the results for a specimen with a double slip orientation on the [100]–[111] line in stereographic projection with the results for a single slip orientation, which was already shown in Fig. 5.5. As shown in Fig. 5.6a, the load-elongation curve of the double-slip orientation specimen shows a stage II to III deformation curve, and the plastic flow stress is considerably higher than that of the single-slip orientation specimen. When the tensile axis is oriented for double slip, the dislocations on the two slip systems are arranged such that they become forest dislocations with respect to each other, and the mean free path of the dislocations decreases rapidly according to Eq. (4.13). The result for the dislocation mean free path obtained during the analysis is shown in Fig. 5.6b. Figure 5.6c shows the evolutions of the vacancy density. These results show that the vacancy density is significantly higher when multiple slip deformation occurs compared to the case of single slip deformation.

Table 5.1 Input data used for the analysis for the development of atomic vacancies in a FCC single crystal under monotonic loading

Description	Input data
Elastic compliance	$s_{11} = 1.4995$, $s_{12} = -0.6282$, $s_{44} = 1.3263 \times 10^{-11}$ Pa^{-1}
Magnitude of Burgers vector	$b = 2.556 \times 10^{-10}$ m
Coefficient for the Taylor hardening term	$a = 0.1$
Aspect ratio of dislocation loop	$\alpha = 1.0$
Annihilation distance	$D = 5b = 1.278 \times 10^{-9}$ m
Model for the CRSS	Equation (4.15)
Type of the dislocation mean free path model	Equation (4.22)
Parameters for the microstructure representative length	$d = 1000$ μm, $c_T = 3.0$, $\beta = 1.0$
Lattice friction stress	$\theta_0 = 0.3$ MPa
Parameters for the dislocation mean free path model	$c^* = 100.0$, $n^* = 1.0$
Parameters used in atomic vacancy evolution model	$c_e = 0.5$, $y_p = b$
Total initial dislocation density on twelve slip systems	$\rho_0 = 1.2 \times 10^{10}$ m^{-2}

In polycrystals or in multi-phase materials, slip deformation on multiple slip systems often occurs near grain or phase boundaries in order to satisfy compatibility of deformation between adjacent grains or regions of different phase. The results in Fig. 5.6 show that when multiple slip deformation occurs near the grain or phase boundaries in the microstructure, a rapid increase in vacancy density also occurs. One might further imagine that a large amount of vacancies would diffuse to nearby grain or phase boundaries, or junctions of boundaries, such as grain boundary triple lines or quadruple points. High density of vacancies thus collected will eventually form nanometer-sized voids and grow further if the atomic vacancies are supplied continuously. When voids of larger size, such as shown in Fig. 5.1, are formed and connected, embryos of microcrack will be generated.

The amount of vacancies swept by the dislocation has already been explained in Eq. (5.9). In principle, when a vacancy is absorbed by an edge dislocation, an atomic-scale jog is formed. If the number of vacancies absorbed is large, the mobility of the dislocations is considered to be reduced. The resistance to slip deformation resulting from this effect would be a function of the distance between vacancies. We already assumed that vacancies in the region of thickness $2y_p$ above and below the slip plane are swept out by

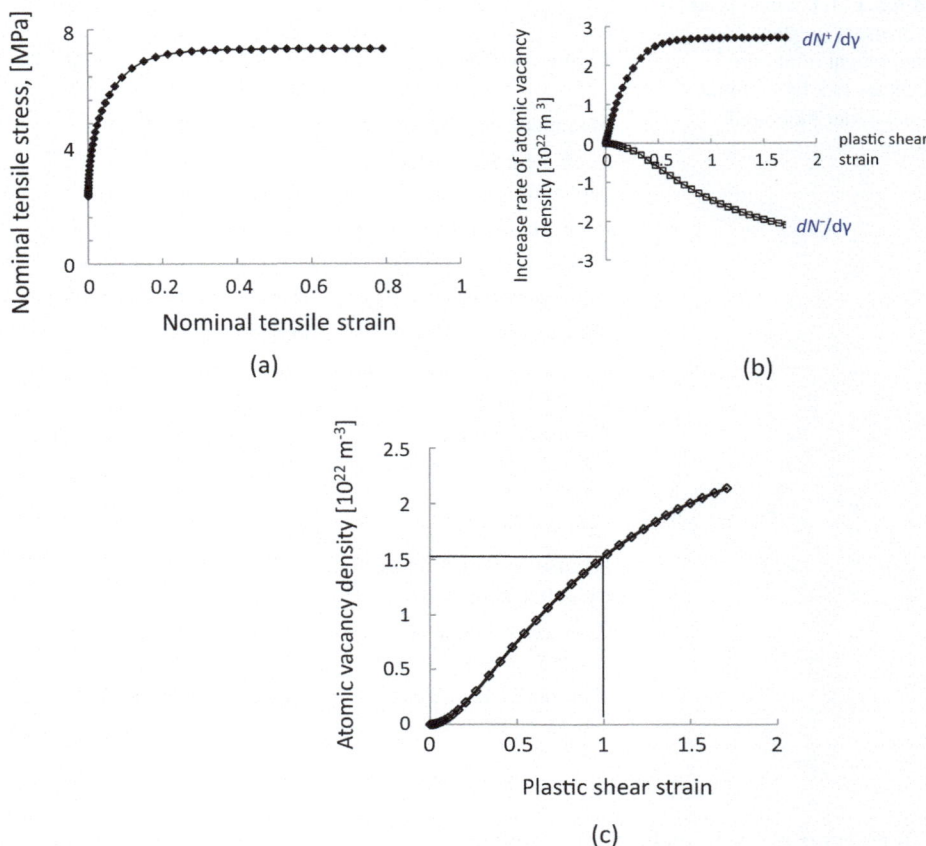

Fig. 5.5 Evolution of atomic vacancy density in Cu single crystals under monotonic tensile loading. Input data used for the analysis are given in Table 5.1. **a** load-elongation curve, **b** rates of vacancy generation and sweep, **c** evolution of vacancy density

the movement of a dislocation. Then in turn, we assume that these vacancies affect the movement of the dislocation. The average distance of vacancies in this region $\bar{l}_{vac}$ is,

$$\bar{l}_{vac} = 1/\sqrt{2Ny_p}. \tag{5.15}$$

Assuming that the resistance from vacancies to the movement of dislocations is proportional to $1/\bar{l}_{vac}$, we may add a contribution to the CRSS of Eq. (4.15) as,

$$\theta^{(n)} = \theta_0^{(n)}(T) + \sum_{m=1}^{N} \Omega^{(nm)} a\mu b^{(m)} \sqrt{\rho_S^{(m)}} + 3\beta \frac{\mu b}{d^{(n)}} + k_{vac}\mu b^{(n)} \sqrt{2Ny_p}, \tag{5.16}$$

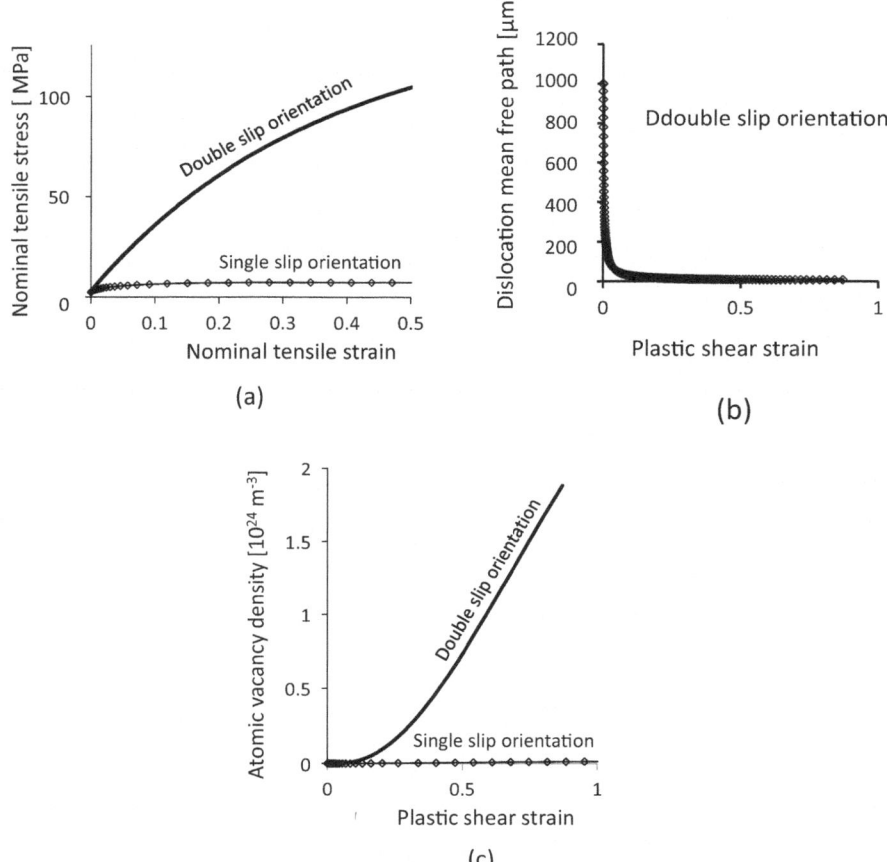

Fig. 5.6 Response of a specimen with double slip orientation. Input data, except for the crystal orientation, are the same as those used for the analysis shown in Fig. 5.5. **a** Load-elongation curve, **b** dislocation mean free path, **c** vacancy density

where, the fourth term on the right hand side is the contribution from the absorbed vacancies and k_{vac} is a numerical coefficient. Quantitative discussions of k_{vac} are beyond the scope of this article, but k_{vac} will depend on strain rate and temperature.

References

1. Hull D, Bacon DJ (2001) Introduction to Dislocations, 4th edn. Butterworth Heinemann, Oxford
2. Argon AS (2008) Strengthening Mechanisms in Crystal Plasticity, Oxford Uni. Oxford University Press, Oxford
3. Takamura J, Takahashi I, Amano M (1969) Iron Steel Inst Japan-Trans 9:216
4. Kimura H, Maddin R (1964) Acta Metall 12:1065
5. Trinkle DR (2017) Philos Mag 97:2514
6. Ozawa E, Kimura H (1970) Acta Metall 18:995
7. Feller-Kniepmeier M, Gobrecht J (1971) Acta Metall 19:569
8. Nabarro FRN (1990) Acta Metall Mater 38:637
9. Cuitiño AM, Ortiz M (1996) Acta Mater 44:427
10. Goods SH, Brown LM (1979) Acta Metall 27:1
11. Hsiung LM, Stoloff NS (1990) Acta Metall Mater 38:1191
12. Polák J, Man J (2014) Int J Fatigue 65:18
13. Antolovich SD, Armstrong RW (2014) Prog Mater Sci 59:1
14. Nagumo M, Yagi T, Saitoh H (2000) Acta Mater 48:943
15. Essmann U, Mughrabi H (1979) Philos Mag A 40:731
16. Kimura H (1998) Concepts of the Strength of Materials (in Japanese), 1st editi. Agne Publishing, Tokyo
17. Ohashi T, Okuyama Y (2019) Philos Mag 99:3032
18. Ohashi T (2018) Philos Mag 98:2275

Mathematical Framework of Crystal Plasticity Analysis Constructed by Hill and Implementation of Dislocation Models

6

Abstract

The constitutive equation of the crystal plasticity analysis constructed by Hill is presented and the hardening matrix used in the equation is quantified by the models for the behavior of dislocations.

Keywords

Schmid tensor · Increment of strain · Generalized Hook's law · Hardening matrix

Mathematical framework of the crystal plasticity analysis constructed by Hill [1] is as follows. The activation condition of slip systems is given by Schmid's law. That is, the resolved shear stress (RSS) acting on a slip system must reach the critical resolved shear stress (CRSS), and at the same time increment of the RSS should be equal to the increment of CRSS.

$$P_{ij}^{(n)}\sigma_{ij} = \theta^{(n)}, \tag{6.1}$$

$$P_{ij}^{(n)}\dot{\sigma}_{ij} = \dot{\theta}^{(n)}, \tag{6.2}$$

where, $P_{ij}^{(n)}$ denotes the Schmid tensor of the slip system n and is defined by the unit vector $\mathbf{v}$ normal to the slip plane and the unit vector $\mathbf{b}$ parallel to the slip direction,

$$P_{ij}^{(n)} = \frac{1}{2}\left\{v_i^{(n)}b_j^{(n)} + v_j^{(n)}b_i^{(n)}\right\}. \tag{6.3}$$

The increment of strain is the sum of elastic and plastic components,

$$\dot{\varepsilon}_{ij} = \dot{\varepsilon}_{ij}^{e} + \dot{\varepsilon}_{ij}^{p}. \tag{6.4}$$

© The Author(s), under exclusive license to Springer Nature Switzerland AG 2024
T. Ohashi, *Mathematical Modeling of Dislocation Behavior and Its Application to Crystal Plasticity Analysis*, Synthesis Lectures on Mechanical Engineering,
https://doi.org/10.1007/978-3-031-37893-5_6

The elastic component is connected to the stress component by the generalized Hook's law, and the plastic component is the sum of contributions from plastic shear strain on slip systems,

$$\begin{cases} \dot{\varepsilon}^e_{ij} = S^e_{ijkl}\dot{\sigma}_{kl} \\ \dot{\varepsilon}^p_{ij} = \sum_n \dot{\gamma}^{(n)} P^{(n)}_{ij} \end{cases}. \tag{6.5}$$

Assuming that the increment of CRSS is defined by the increment of the plastic shear strain as follows,

$$\dot{\theta}^{(n)} = \sum_m h^{(nm)} \dot{\gamma}^{(m)}, \tag{6.6}$$

the relation for the increments of stress and strain components in crystal coordinate system is,

$$\dot{\sigma}_{ij} = \left[S^e_{ijkl} + \sum_n \sum_m \left\{ h^{(nm)} \right\}^{-1} P^{(n)}_{ij} P^{(m)}_{kl} \right]^{-1} \dot{\varepsilon}_{kl}. \tag{6.7}$$

Assuming that CRSS is given by Eq. (4.15) as a function of dislocation density, and that the evolution of dislocation density is obtained by Eq. (3.10), the hardening matrix is obtained as follows,

$$h^{(nm)} = \frac{1}{2}\Omega^{(nm)} a\mu \frac{1}{\sqrt{\rho^{(m)}_s}} \left[\frac{c}{L^{(m)}} - D^{(m)} \rho^{(m)}_s \right], \tag{6.8}$$

where, $c = (1+\alpha)^2/2\alpha$ as shown in Eq. (3.13).

When the dislocation density reaches the following equilibrium value,

$$\rho^{(m)}_{S,eq.} = \frac{c}{D^{(m)} L^{(m)}}, \tag{6.9}$$

then, $h^{(nm)} = 0$, and the strain hardening of the slip system is lost.

Reference

1. Hill R (1966) J Mech Phys Solids 14:95

7

An Example of Analysis: Tension–Compression Straining of a FCC Single Crystal Plate and Bauschinger Effect

Abstract

As an example of crystal plasticity finite element analysis, mechanical response of a single crystal plate with a dislocation cell structure under cyclic loading was examined in detail. The employed model consists of cell walls of 2 μm thickness where the initial density of dislocations is 1.2×10^{16} /m^2, and the interior of cell, whose size is 8 μm and the initial density of dislocations is 1.2×10^{10} /m^2. Macroscopic load-elongation curve obtained by the analysis shows a pronounced Bauschinger effect, which is attributed to stress partitioning taking place in the cell walls and interior. Evolutions of dislocation mean free path, density of atomic vacancy, plastic work density, and cumulative equivalent plastic strain were investigated.

Keywords

Dislocation cell structure · Cyclic loading · Bauschinger effect · Stress partitioning · Plastic work density · Cumulative equivalent plastic strain

In plastic deformation of crystalline materials mediated by movement of dislocations, characteristic patterns sometimes develop in the density distribution of accumulated dislocations. Cell-, vein-, or ladder-structures are well known. At least within the scope of the theory described up to Chap. 6, it is difficult to reproduce the process of formation of these substructure of dislocations by crystal plasticity analysis; however, it is possible to trace the mechanical response of materials in which the substructure of dislocations is formed.

The cell structure consists of a wall-like region with high density of dislocation accumulation, called the cell wall, and cell interior with a low dislocation density. Figure 7.1a shows a schematic illustration of the cell structure by Mughrabi [1], where slip deformation occurs inside the cell and the dislocations causing slip are absorbed into the cell

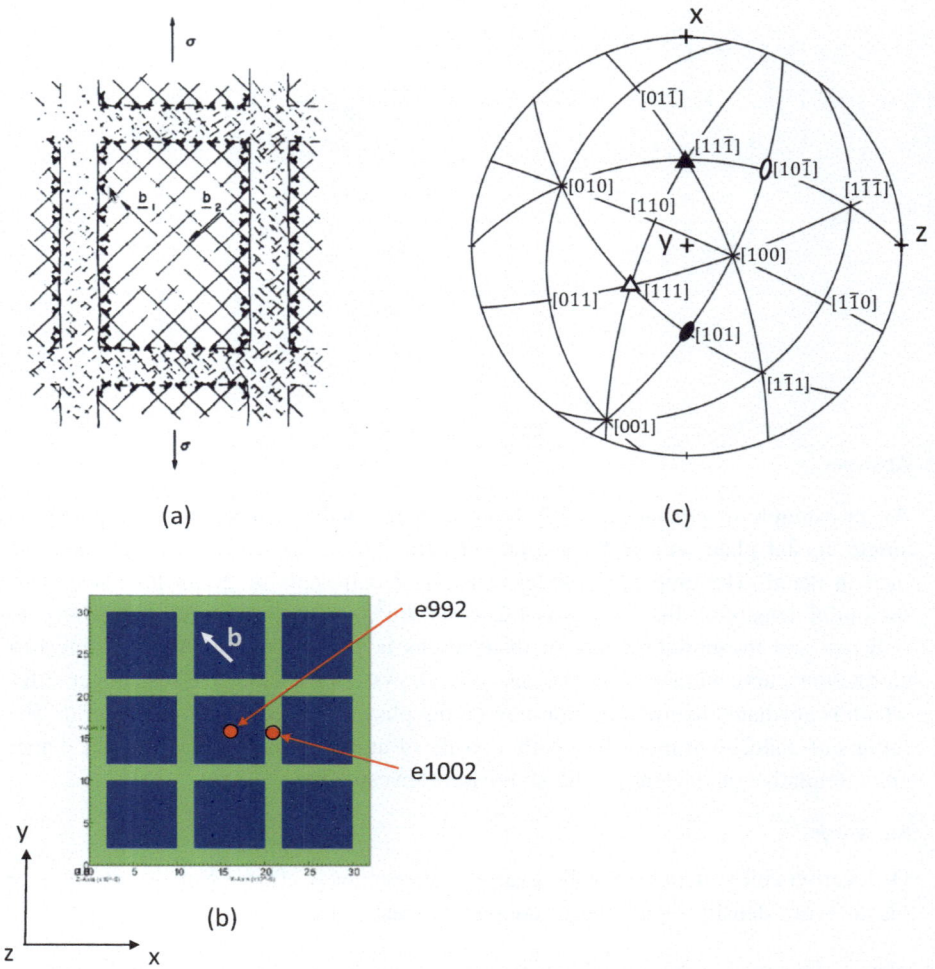

Fig. 7.1 Mechanical response of a single crystal with dislocation substructure model of Mughrabi under tensile and compressive loading. **a** Schematic of dislocation cell structure by Mughrabi [1], **b** model employed for numerical analysis, **c** crystal orientation

walls. Morita and Kuroda [2] discussed the mechanical response of the materials with cell structure using a numerical model that mimics Mughrabi's model. Here, following Mughrabi's model and Morita and Kuroda's approach, we created the numerical model specimen shown in Fig. 7.1b.

The specimen is a rectangular shaped single crystal plate of FCC type with dimensions of $32 \times 32 \times 2$ μm. The specimen was divided into $32 \times 32 \times 2 = 2048$ cube elements.

The initial dislocation density in the green colored wall-like region (i.e. cell walls) is 1×10^{15}/m^2 in twelve {111}<110> slip systems and the total density is 1.2×10^{16}/m^2. The initial dislocation density in the blue colored region (i.e. cell interior) surrounded by cell walls is 1×10^9/m^2 in twelve slip systems and the total density is 1.2×10^{10}/m^2.

The displacement in y direction of the bottom surface of the specimen was constrained, while the top surface was subjected to uniform axial displacement in y direction. Forced displacement is ± 0.96 μm and the nominal axial strain is $\pm 3\%$. Figure 7.1c shows the crystal orientation of the loading axis, which is the same inside the cell and in the cell wall. The crystal orientation was determined to make the slip plane normal direction and slip direction of primary slip system of B4: $(11\bar{1})[101]$, shown in Fig. 7.1c by filled symbols of triangle and ellipse, are parallel to x–y plane. Handling of crystal orientation and stereographic presentation is given in Appendix A. The Schmid factor of the primary system is 0.5. The slip system with the next largest Schmid factor is A3: $(111)[10\bar{1}]$ and the slip plane and slip direction are shown by open symbols in Fig. 7.1c. The Schmid factor of the A3 system is approximately 0.467.

The model of CRSS is given by Eq. (4.15). Details of the input data are shown in Table 7.1. The interaction matrix of slip systems used in the Taylor hardening term is given by $\Omega^{(ii)} = 1$ for the diagonal components and $\Omega^{(ij)} = 1 + \delta$ for the off-diagonal components. A perturbation δ of the order of 0.01 is introduced to suppress numerical instability during the analysis, and the strain hardening of twelve slip systems is essentially isotropic. Equation (4.22) is used for the dislocation mean free path. The length scales of the cell interior and cell wall regions are assumed to be equal to the width of the cell and the wall thickness, respectively. The self- and coplanar-interaction components of the weight matrix $w_S^{(nm)}$ are assumed to be 0 and the other components 1. The same as for $w_S^{(nm)}$ was used for the weight matrix $w_G^{(nm)}$.

Figure 7.2 shows the relationship between nominal stress and nominal strain obtained from the analysis. The red colored broken line indicates elastic relationship, and the steps in the process of cyclic deformation are indicated by #21, #41, etc. Plastic deformation begins below a nominal stress of 100 MPa. The deformation step when the nominal axial strain first reached 3% is #21, and the nominal stress at #21 is approximately 700 MPa. Before the deformation reaches #21, the macroscopic strain hardening rate changes largely when the nominal strain is about 1.5%. This is because plastic slip first occur in the interior of cells, and then the slip occurs also in cell walls. Let us examine this process in detail.

Figure 7.3 shows the mean free paths of dislocations obtained for the finite elements indicated by e992 and e1002 in Fig. 7.1b. The mean free path at the element e992 positioned inside a cell is initially 8 μm and decreases rapidly with deformation to 0.4 μm at #21. The mean free path at the element e1002 positioned in the cell wall is shown by the red curve. The initial dislocation density in the cell wall is very high at 1.2×10^{16}/m^2 and the initial value of the mean free path is approximately 0.1 μm. When the nominal strain is 1.5%, plastic slip deformation starts in the cell walls, which results in an increase in

Table 7.1 Input data used for the analysis of FCC single crystal with dislocation cell structure and subjected to cyclic loading

Description	Input data
Elastic compliance	$s_{11} = 1.4995$, $s_{12} = -0.6282$, $s_{44} = 1.3263 \times 10^{-11}$ Pa^{-1}
Magnitude of Burgers vector	$b = 2.556 \times 10^{-10}$ m
Coefficient for the Taylor hardening term	$a = 0.1$
Aspect ratio of dislocation loop	$\alpha = 1.0$
Annihilation distance	$D = 1.278 \times 10^{-9}$ m
Model for the CRSS	Equation (4.15)
Type of the dislocation mean free path model	Equation (4.22)
Parameters for the microstructure representative length	$d = 8$ μm (cell interior), $d = 2$ μm (cell wall), $c_T = 3.0$, $\beta = 1.0$
Lattice friction stress	$\theta_0 = 30$ MPa
Parameters for the dislocation mean free path model	$c^* = 10.0$, $n^* = 1.0$
Parameters used in atomic vacancy evolution model	$c_e = 0.5$, $y_p = b$
Contribution coefficient of atomic vacancies to CRSS	$k_{vac} = 0$

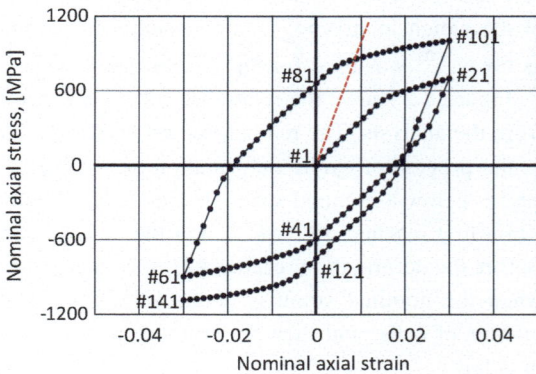

Fig. 7.2 Load-elongation curve of a Cu single-crystal specimen with dislocation cell structure. The elastic line of the material is indicated by the red dashed line. A pronounced Bauschinger effect occurs during load reversal

SS dislocation density and a gradual decrease in the mean free path. At #21, the mean free path in e1002 was about 0.09 μm.

The strain hardening and the change of its rate occurred between #1 and #21 shown in Fig. 7.2 are caused by the rapid decrease in the mean free path and the propagation of slip deformation to the cell walls. As the applied load is reduced, the elements e992 and e1002 recover elastically until the nominal strain is about 2.2%, and the mean free paths

Fig. 7.3 Variation of the mean free path of dislocations on the primary slip system in elements e992 and e1002

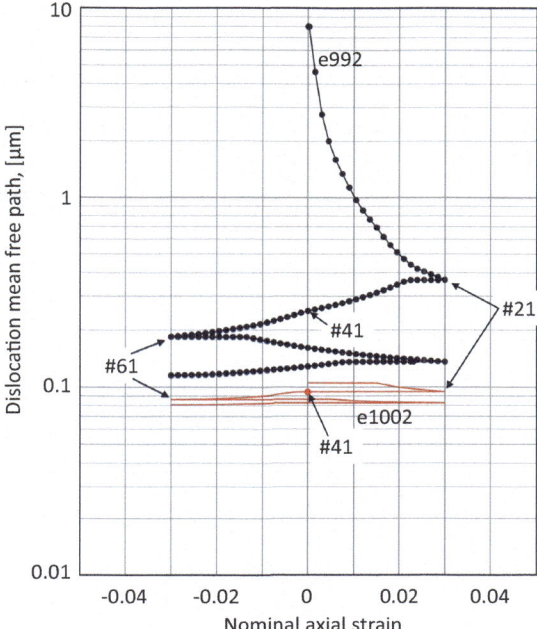

are kept constant. Thereafter until #41, slip deformation occurs in the cell interior, and from #41 to #61, slip deformation also occurs in the cell walls. A significant Bauschinger effect is evident between #21 and #61.

Figure 7.4 shows the distribution of the normal stress in the loading direction. At #21 when the first tensile deformation ends, a large tensile stress is developed in the region of cell walls, showing that a stress partitioning is established in the microstructure. The plastic slip deformation in the cell walls is delayed because the density of SS dislocation is high and the CRSS is larger, and cell walls support a larger part of applied load compared to the cell interior where slip deformation occurred earlier. At the point where the applied load decreases, cell walls are in a state of large elastic tension and act to compress the cell interior, causing slip deformation in the cell interior at the nominal strain of about 2.2%, as shown in Figs. 7.2 and 7.3. This slip is in the opposite direction to that occurred during #1 and #21.

When the applied load turn to compression, plastic relaxation of cell walls is delayed compared to the cell interior, resulting in a larger compressive stress state as shown in Fig. 7.4b and c. Figure 7.4d shows the stress distribution at #101 when the second tensile deformation ends. A distribution similar to that observed in Fig. 7.4a is formed. That is, in materials with dislocation cell structures, a large stress partitioning occurs between the cell walls and the interior and this causes the Bauschinger effect.

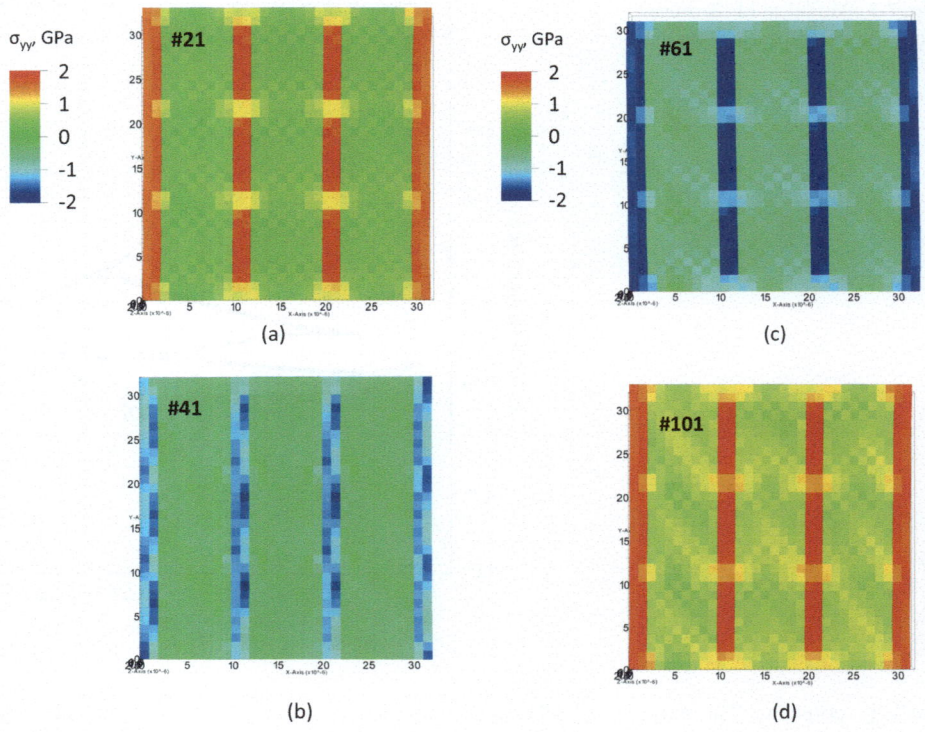

Fig. 7.4 Distribution of the normal stress σ_{yy} in the loading direction at deformation stages **a** #21, **b** #41, **c** #61 and **d** #101. (Visualization: VisIt 3.1.4, Lawrence Livermore National Security, LLC)

Figure 7.5 shows the distribution of the plastic work density done until #101 by the primary and the conjugate slip systems of B4 and A3. (Refer to the Appendix B for the evaluation of the plastic work density and the cumulative plastic strain.) As described before, the Schmid factors of the primary and the conjugate systems are close to each other and 0.5 and 0.467, respectively. The plastic work density done by the primary system is large in the cell walls, while the density done by the conjugate system is large inside the cells.

How the dominant slip system is selected in the cell walls and in the cell interiors is not simply determined by the Schmid factor. This is because the deformation in the cell structure is largely non-uniform and the stress field is inherently tri-axial even if the applied load is uniaxial. Inside the cells surrounded by stiff walls, activity of only one slip system results in an anisotropic deformation and induce activity of other slip systems to mitigate the anisotropy or to maintain the compatibility of deformation with the cell-wall regions. In other words, slip activity in the cell interior is not determined solely from the external loading condition and crystal orientation. Plastic deformation also occurs in the

Fig. 7.5 Distribution of plastic work density done by **a** B4, and **b** A3 slip systems by the step #101. Slip deformation occurs not only inside the cell but also in the cell walls

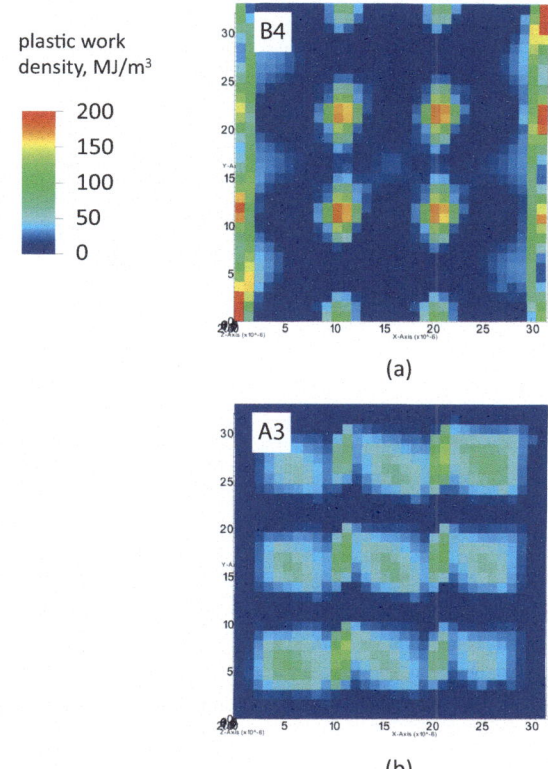

cell wall, indicating that the cell wall is not a "frozen" structure and may dynamically change its spatial configuration and size in response to external loading.

Figure 7.6 shows the distribution of the cumulative equivalent plastic strain. The strain in the cell wall is about 1/4 that of the cell interior. As already shown in Fig. 7.5, the plastic work density was larger in the cell walls, while the cumulative equivalent strain is larger in the cell interior. This difference reflects the difference in the magnitude of stress occurring inside the cell and in the cell wall.

Figure 7.7 shows the density evolution of atomic vacancies. At #21, the vacancy density is 10^{23} to $10^{24}/m^3$, and the density increases with repeated deformation, reaching $10^{25}/m^3$ both in the cell interior and cell walls at #141.

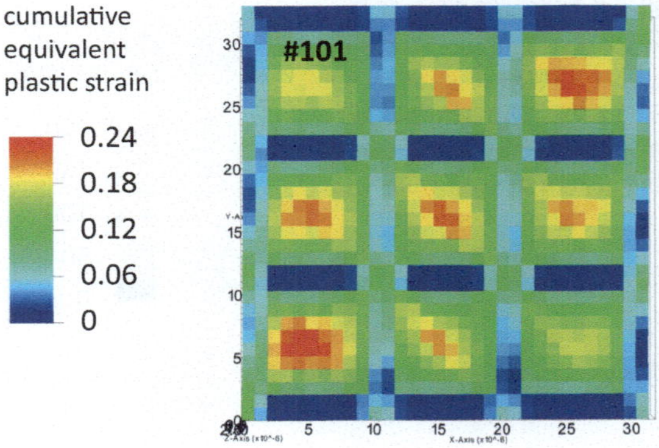

Fig. 7.6 Cumulative equivalent plastic strain at deformation stage #101

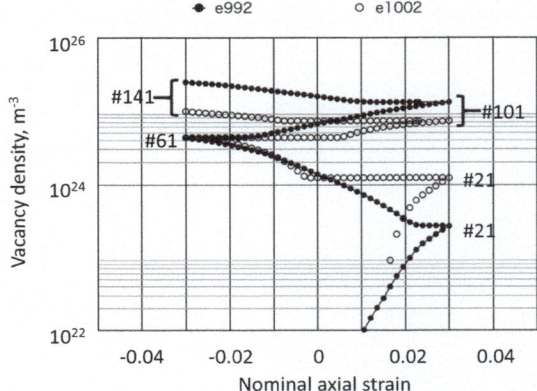

Fig. 7.7 Evolution of the atomic vacancy density inside the cell (e992) and cell wall (e1002). The atomic vacancy density is on the order of 1025/m^3 at the #141 deformation stage. Since the atomic number density is about 8.5×1028/m^3, the vacancy density in terms of concentration is on the order of 10^{-4}

References

1. Mughrabi H (1983) Acta Metall 31:1367
2. Morita S, Kuroda M (2018) Investigation of the origins of Bauschinger effect in polycrystalline metals. In: Proc. JSME 31 St Comput. Mech. Devision Conf. The Japan Society of Mechanical Engineers, Tokushima

Appendix A: Coordinate Transformation by Euler Angles and Graphical Presentation by Pole Figures

The coordinate transformation matrices for rotating the coordinates around the first, second or third coordinate axis are as follows,

$$rot1(\alpha) = \begin{pmatrix} 1 & 0 & 0 \\ 0 & \cos\alpha & \sin\alpha \\ 0 & -\sin\alpha & \cos\alpha \end{pmatrix},$$

$$rot2(\alpha) = \begin{pmatrix} \cos\alpha & 0 & -\sin\alpha \\ 0 & 1 & 0 \\ \sin\alpha & 0 & \cos\alpha \end{pmatrix},$$

$$rot3(\alpha) = \begin{pmatrix} \cos\alpha & \sin\alpha & 0 \\ -\sin\alpha & \cos\alpha & 0 \\ 0 & 0 & 1 \end{pmatrix}.$$

Suppose that the coordinate axis parallel to the [100], [010] and [001] directions of the crystal, which make up the crystal coordinate system, initially coincides with the x–y–z coordinate system as shown in Fig. A.1a. When the crystal coordinate system is rotated φ with [001] as the rotation axis as shown in (b), and then by θ with [100] as the rotation axis as shown in (c), and finally by κ with [001] as the rotation axis as shown in (d), the relationship between the crystal and x–y–z coordinate systems are as follows,

$$\begin{pmatrix} [100] \\ [010] \\ [001] \end{pmatrix} = rot3(\kappa)\,rot1(\theta)\,rot3(\varphi) \begin{pmatrix} x \\ y \\ z \end{pmatrix}$$

$$= \begin{pmatrix} -\sin\kappa\cos\theta\sin\phi + \cos\kappa\cos\phi & \sin\kappa\cos\theta\cos\phi + \cos\kappa\sin\phi & \sin\kappa\sin\theta \\ -\cos\kappa\cos\theta\sin\phi - \sin\kappa\cos\phi & \cos\kappa\cos\theta\cos\phi - \sin\kappa\sin\phi & \cos\kappa\sin\theta \\ \sin\theta\sin\phi & -\sin\theta\cos\phi & \cos\theta \end{pmatrix} \begin{pmatrix} x \\ y \\ z \end{pmatrix}$$

(A.1)

© The Editor(s) (if applicable) and The Author(s), under exclusive license to Springer Nature Switzerland AG 2024
T. Ohashi, *Mathematical Modeling of Dislocation Behavior and Its Application to Crystal Plasticity Analysis*, Synthesis Lectures on Mechanical Engineering, https://doi.org/10.1007/978-3-031-37893-5

Fig. A.1 Rotation of the crystal coordinate system by Euler angles

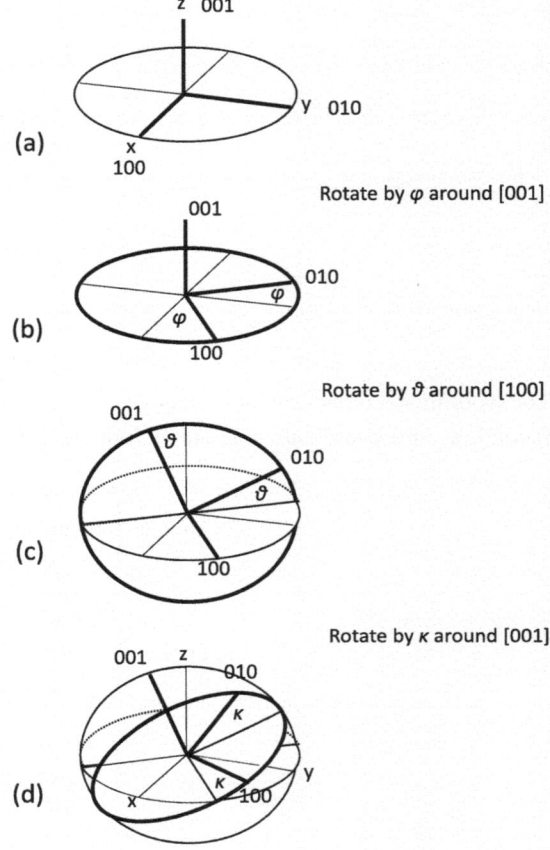

Since this coordinate transformation matrix is orthonormal, the inverse of (A.1) is equal to its transposed matrix,

$$\begin{pmatrix} x \\ y \\ z \end{pmatrix} = \begin{pmatrix} -\sin\kappa \cos\theta \sin\phi + \cos\kappa \cos\phi & -\cos\kappa \cos\theta \sin\phi - \sin\kappa \cos\phi & \sin\theta \sin\phi \\ \sin\kappa \cos\theta \cos\phi + \cos\kappa \sin\phi & \cos\kappa \cos\theta \cos\phi - \sin\kappa \sin\phi & -\sin\theta \cos\phi \\ \sin\kappa \sin\theta & \cos\kappa \sin\theta & \cos\theta \end{pmatrix} \begin{pmatrix} [100] \\ [010] \\ [001] \end{pmatrix}$$
(A.2)

The rotation angles, κ, θ, and φ are called Euler angles and are widely used to define the rotation relationship between the crystal coordinate system and the specimen coordinate system, which is defined with respect to external conditions such as the direction of applied load or the external shape of the specimen. As one might imagine, there are arbitrariness in the order of rotation and the choice of rotation axes, and it is possible to define a coordinate transformation matrix for each case. Bunge angle is one such example.

Fig. A.2 Stereo projection of crystal orientations when the Euler angles are $\kappa = \theta = \varphi = 0$ and the projection pole is z. Orientations [100], [010], and [001] coincide to x, y, and z axis, respectively

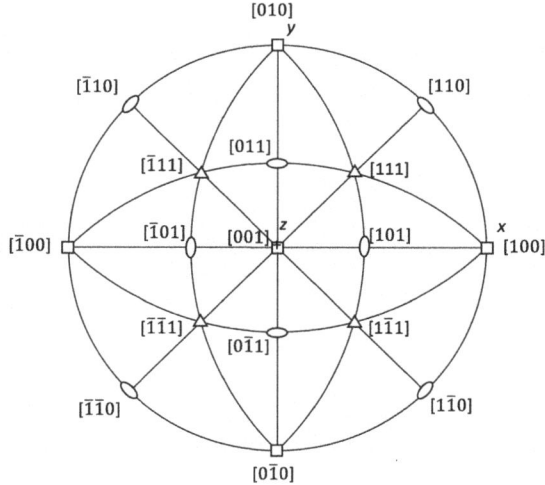

Pole figure has long been used as a method of displaying crystal orientation in two-dimensional space. Figure A.2 shows the [001] pole figure. Directions of x, y, and z axes are also shown when the Euler angle is $\kappa = \theta = \varphi = 0$, which coincide with [100], [010], and [001].

As an example, Fig. A.3a shows the pole figure when $\kappa = 77°$, $\theta = 24.7°$, $\varphi = 257.6°$ with the z-axis as the pole. Consider the case of FCC crystals. When uniaxial tension or compression in the z-axis is applied, the slip system with the largest Schmid factor among {111}<110> slip systems is C5: $(1\bar{1}1)[011]$, the slip plane is $(1\bar{1}1)$ and the slip direction is [011]. As shown in Fig. A.3a, the x-component is zero for the direction of the slip plane normal and slip direction in this orientation. Figure A.3b schematically shows the arrangement of this slip system, assuming a rectangular shaped single crystal.

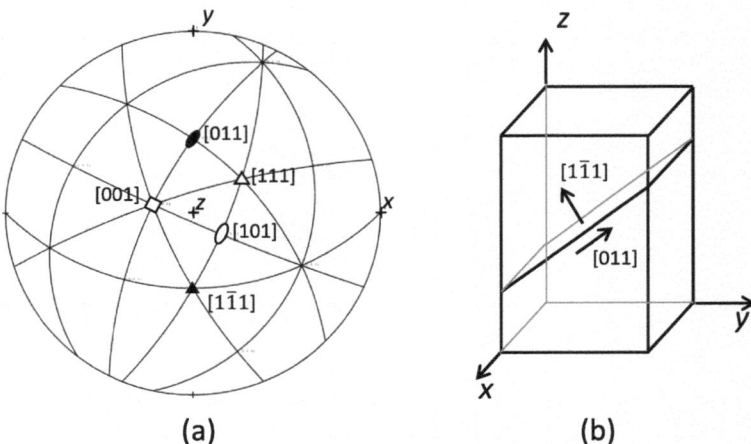

Fig. A.3 **a** Stereo projection when $\kappa = 77°$, $\theta = 24.7°$, $\varphi = 257.6°$. In this case, the z-axis is in a triangle composed of three orientations [001], [101], and [111]. Orientations $[1\bar{1}1]$ and [011] are both parallel to the y–z plane and have no x component. **b** Schematic diagram showing the arrangement of slip plane $(1\bar{1}1)$ and slip direction [011] in a rectangular single crystal specimen when the crystal orientation is that shown in (a). When an axial load in z-direction is applied to this specimen the maximum resolved shear stress among twelve {111}<110> slip systems occurs in this system and the slip deformation is parallel to the y–z plane

Appendix B: Work Done by Slip Deformation and Equivalent Plastic Strain

Let us summarize the quantity that represents the overall magnitude of plastic deformation that has occured in the crystal. The incremet of work done to the material by plastic slip is the sum of the increment of work done in each slip system,

$$dW^p = \sum_n \tau^{(n)} \cdot d\gamma^{(n)}, \tag{B.1}$$

where, τ and $d\gamma$ are the resolved shear stress (RSS) on the slip system and the increment of the shear strain, respectively. Since the unit of RSS is N/m^2, the unit of the work increment dW^p is Nm/m^3 = J/m^3, meaning that dW^p is the spatial density of plastic work increment.

It is sometimes useful to define some scalar quantity that represents the magnitude of load when a material is subjected to multiaxial load or when a tri-axial stress state is formed in the microstructure. One such scalar quantity is the equivalent stress, defined by the six stress components or three principal stresses $\sigma_1, \sigma_2, \sigma_3$ as follows,

$$\begin{aligned}\sigma_{eq} &= \sqrt{\frac{1}{2}\left\{(\sigma_{xx}-\sigma_{yy})^2 + (\sigma_{yy}-\sigma_{zz})^2 + (\sigma_{zz}-\sigma_{xx})^2 + 6\left(\sigma_{xy}^2 + \sigma_{yz}^2 + \sigma_{zx}^2\right)\right\}} \\ &= \sqrt{\frac{1}{2}\left\{(\sigma_1-\sigma_2)^2 + (\sigma_2-\sigma_3)^2 + (\sigma_3-\sigma_1)^2\right\}}\end{aligned} \tag{B.2}$$

Assuming that the increment of plastic work density dW^p is given by the product of equivalent stress and a certain scalar quantity $d\varepsilon_{eq}^p$, we write,

$$dW^p = \sigma_{eq} \cdot d\varepsilon_{eq}^p, \tag{B.3}$$

where, $d\varepsilon_{eq}^p$ is called the increment of equivalent plastic strain. From Eqs. (B.1)–(B.3), we obtain,

$$d\varepsilon_{eq}^p = dW^p/\sigma_{eq} = \frac{\sum_n \tau^{(n)} \cdot d\gamma^{(n)}}{\sigma_{eq}}. \tag{B.4}$$

$d\varepsilon_{eq}^{p}$ serves as a scalar measure of the quantity of plastic deformation increment. Integration of $d\varepsilon_{eq}^{p}$,

$$\varepsilon_{eq}^{p} = \int d\varepsilon_{eq}^{p}, \qquad (B.5)$$

is the cumulative equivalent plastic strain and serves as a scalar measure of the overall quantity of plastic deformation in the material.

Index

A

Absorption of dislocations, 55
Accumulation of dislocations, 7
Activation energy for the dislocation cutting, 17
Annihilation distance. *See also* Dislocation annihilation
Annihilation of dislocations, *see* Dislocation annihilation
Asymptotic dislocation density, 11
Athermal and thermal terms, 17
Atomic vacancies, 57
    resistance to slip deformation by atomic vacancy, 63
Atomic vacancies and physical properties, 58

B

Bagaryatsky, 55
Bauschinger effect, 69
BCC crystals, 33
Boltzmann constant, 17, 57
Burgers circuit, *see* Geometrically necessary dislocations

C

Cell interior, 69
Cell structure, 69
Cell wall, 69
Characteristic angle of GN dislocations, *see* Geometrically necessary dislocations
Characteristic length scale, *see* Microstructural length scale
Compatibility of deformation, 74
Contribution from slip systems, 29
Coordinate transformation matrix, 77
Critical resolved shear stress, 10, 25, 49
Cross-slip systems, 22
CRSS, 29. *See also* Critical resolved shear stress
Crystallographically determined twin strain, 50
Cumulative equivalent plastic strain, 75, 82
Cyclic loading, 69

D

Deformation modes, 29
Deformation stages, 14
Deformation stage II, *see* Deformation stages
Deformation twin, 49
Density evolution of SS dislocations, *see* Statistically stored dislocations
Density norm of GN dislocations, *see* Geometrically necessary dislocations
Dependence of the Taylor term on temperature and strain rate, 28
Dependence to temperature and strain rate, 29
Direction of dislocation line, *see* Geometrically necessary dislocations
Dislocation annihilation. *See also* Vacancy formation mechanism
Dislocation cell structure, 69
Dislocation density, 10

Dislocation density tensor, *see* Geometrically necessary dislocations
Dislocation dynamics simulations, 44
Dislocation interactions, 31
Dislocation loops, 7, 20
Dislocation mean free path, 7, 8, 10, 11, 13, 15, 39, 51, 62
Dislocations, 1
Dragging of dislocation jogs, 58
Dynamic recovery, 9

E
Effective mean spacing of dislocations, 52
Emission of dislocation loops, 7, 45
Emission of prismatic dislocation loops, *see* Geometrically necessary dislocations
Equivalent plastic strain, 81
Equivalent stress. *See also* Plastic work density
Euler angles. *See also* Coordinate transformation matrix
Extra half-planes of edge dislocations, *see* Atomic vacancies
Extra half planes of GN dislocations, *see* Geometrically necessary dislocations

F
FCC crystals, 12, 29
Ferrite-cementite interface, 52
Ferrite-cementite phase boundaries, 48
Ferrite lamellae, 48
Finite element method, 1
Formation enthalpy of atomic vacancies, *see* Atomic vacancies
Fracture surface, 58
Frank-Read dislocation sources, 44
Frank-Read source, 44
Frank-Read type dislocation loop emission". , *see* t→ *Frank-Read source*
Frank-Read type dislocation source". , *see* t→ *Frank-Read source*

G
Generalized Hook's law, 68
Geometrically necessary dislocations, 18
GN dislocations, *see* Geometrically necessary dislocations

Gradient in plastic shear strain, *see* Geometrically necessary dislocations
Grain boundaries, 40, 48

H
Half shear loops, *see* Geometrically necessary dislocations
Hardening matrix, 68
HCP crystal lattice, 33
Hydrostatic stress field, 21

I
Increment of dislocation density, 8
Increment of strain, 67
Increment of vacancy density, *see* Atomic vacancies
Initial inhomogeneities, 14
Interaction between dislocations, 31
Interaction matrix, 25, 28, 29, 31, 33, 39, 42
Internal state variable. *See also* State variables
Internal stress field, 26

L
Latent hardening, 33
Lattice friction stresses, 28
Length scale, *see* Microstructural length scale, 50
Length scale of metal microstructure, 25, 40
Linear strain hardening, *see* Deformation stages
Load-elongation curve, 12, 62

M
Materials science of lattice defects, 1
Mathematical framework, 67
Microstructural length scale, 9, 48, 51
Microstructure, 2
Miller-Bravais indices, 33
Model of Seeger et al., 12
Multiple slip, 14
Multiplication of slip on different slip systems, 2

N

Non-uniform initial dislocation density, *see* Initial inhomogeneities
Non-uniform slip deformation, 2

O

Orowan stress term in a finite medium, 48
Overall magnitude of plastic deformation, *see* Equivalent plastic strain

P

Pair annihilation of edge dislocations, 58
Parabolic strain hardening. *See also* deformation stages
Pearlite microstructure, 52
Pearlite steels, 48
Phase boundaries, 48
Pitsch-Petch, 55
Plastic deformation of metallic materials, 1
Plastic work density, 74. *See also* Equivalent plastic strain
Pole figure, 79
Precipitates, 48
Precipitation, 52
Precipitation-hardened steel, 53
Prismatic dislocation loops, 23
Probability of trapping by interacting dislocations, 16

R

Rate of vacancy formation, *see* Atomic vacancies
Rate of vacancy sweep, *see* Atomic vacancies
Representative length of the microstructure, 48

S

Saturation of dislocation density, 62
Scale dependent nature of geometrically necessary dislocations, 25
Scale dependent strain hardening characteristics, 40
Schmid tensor, 67
Secondary particles, 48
Shockley partial dislocations, 31
Slip directions, 29
Slip planes, 29
Slip systems, 10, 29
Solid mechanics, 1
Spatial gradient of plastic shear strain, *see* Geometrically necessary dislocations
profiles of plastic shear strain, 20
Spherical void, 21
SS dislocations, *see* Statistically stored dislocations
Stage I hardening, *see* deformation stages
Stage II to III deformation curve, 62
State variables, 15
Statistically stored dislocations, 7, 9
Stereo projection, 79
Strain hardening, 1, 10
Strain rate, 9
Stress partitioning, 69
Stress-strain curves, 10

T

Tangent vector of dislocation line, *see* Geometrically necessary dislocations
Taylor model, 11, 27
Taylor relation. *See also* Taylor model
Taylor terms, 29
Temperature, 9, 17, 57
Thermal equilibrium concentration of vacancies, 57
Thermal process, 16
Thermal term, 17
Thompson tetrahedron, 29
Transition from stage I to II, *see* Deformation stages
Travel distances of dislocation segments, *see* Dislocation mean free path
Tri-axial stress state, 15
Twin boundaries, 49
Twinning, 35, 49
Twin nucleation, 51
Twin systems, 35

V

Vacancies emitted by pair annihilation of edge dislocations, *see* Atomic vacancies
Vacancies swept out by dislocations, *see* Atomic vacancies

Vacancy formation mechanism, *see* Atomic vacancies
Visualization of GN dislocations. *See also* Geometrically necessary dislocations
Void, 58

W

Weight matrices for SS and GN dislocations, 39

Weight matrix of dislocation interaction. *See also* Weight matrices for SS and GN dislocations

Y

Yielding, 1, 42
Yield stress, 42

SPRINGER NATURE

GPSR Compliance

The European Union's (EU) General Product Safety Regulation (GPSR) is a set of rules that requires consumer products to be safe and our obligations to ensure this.

If you have any concerns about our products, you can contact us on ProductSafety@springernature.com

In case Publisher is established outside the EU, the EU authorized representative is:

Springer Nature Customer Service Center GmbH
Europaplatz 3
69115 Heidelberg, Germany

The manufacturer's authorised representative in the EU is Springer Nature Customer Service Centre GmbH, Europaplatz 3, 69115 Heidelberg, Germany. If you have any concerns regarding our products, please contact ProductSafety@springernature.com

Printed and bound by CPI Group (UK) Ltd, Croydon, CR0 4YY

26/03/2026

02078977-0005